History of Algebraic Geometry

The Wadsworth Mathematics Series

W. Beckner, A Calderón, R. Fefferman, P. Jones, *Conference on Harmonic Analysis in Honor of Antoni Zygmund*
M. Behzad, G. Chartrand, L. Lesniak-Foster, *Graphs and Digraphs*
J. Cochran, *Applied Mathematics: Principles, Techniques, and Applications*
W. Derrick, *Complex Analysis and Applications*, Second Edition
J. Dieudonné, *History of Algebraic Geometry*
R. Durrett, *Brownian Motion and Martingales in Analysis*
A. Garsia, *Topics in Almost Everywhere Convergence*
K. Stromberg, *An Introduction to Classical Real Analysis*
R. Salem, *Algebraic Numbers and Fourier Analysis*, and L. Carleson, *Selected Problems on Exceptional Sets*

History of Algebraic Geometry

An Outline of the History and Development of Algebraic Geometry

JEAN DIEUDONNÉ

Translated by Judith D. Sally
Northwestern University

WADSWORTH ADVANCED BOOKS & SOFTWARE
MONTEREY, CALIFORNIA
A DIVISION OF WADSWORTH, INC.

Acquisitions Editor: John Kimmel
Production Editor: Marta Kongsle
Copy Editor: Marion Hanson
Designer: Andrea Cava

Originally published as *Cours de géométrie algébrique I*

Printed in the United States of America

1 2 3 4 5 6 7 8 9 10—89 88 87 86 85

Library of Congress Cataloging in Publication Data

Dieudonné, Jean Alexandre, 1906–
History of algebraic geometry.

Translation of: Cours de géométrie algébrique.
Bibliography: p.
Includes index.
1. Geometry, Algebraic—History. I. Title.
QA564.D513 1985 516.3′5 84-17213
ISBN 0-534-03723-2

ISBN 0-534-03723-2

CONTENTS

Foreword

Algebraic geometry is undoubtedly the area of mathematics where the deviation is greatest between the intuitive ideas forming its starting point and the abstract and complex concepts at the foundation of modern research. The purpose of this work is to help the willing reader bridge the gap.

In this volume, an attempt is made to show how geometers have been led progressively to broaden their ideas. The reader will discover here several fundamental ideas that are revived time after time in different guises, marking the incessant effort toward better understanding of geometric phenomena: transformations and correspondences, invariants, "infinitely near" points, extensions of geometric objects by the "adjunction" of new "points." The reader will also see how the arsenal of algebraic geometry is enriched, little by little, with loans from analysis and topology, where powerful tools have been found, and, more recently, from commutative algebra and homological algebra. Right now, algebraic geometry appears as one of the components of a trinity, the other two members being number theory and the theory of analytic spaces (the modern form of the "theory of several complex variables" for which, according to Serre, the name "analytic geometry" is appropriate[1]). The universal tendency of all the research in these three areas (in which, moreover, the theory of Lie groups, differential geometry, and functional analysis, particularly the theory of partial differential equations, get entangled more and more tightly) is toward a fusion of their basic concepts and of their fundamental methods, so much so that it becomes more and more artificial to dissociate them. Unfortunately, the frame of this work does not permit a better illumination of this gradual fusion, and the author apologizes for giving a deformed image of the truth in describing the influences exercised on algebraic geometry by her neighbors and not the inverse influences that are just as important; the reader must be content, in this regard, with several indications in the bibliography.

To conform with the spirit of this collection, knowledge of mathematical notions beyond the level of the first cycle at the university is not assumed from the start; the new concepts, introduced in algebraic geometry up to 1940, which are above this level, have been described succinctly as the text progresses. These descriptions are supplemented by references in the annotated bibliography at the

[1] There is no reason to keep the old appellation of "analytic geometry" to designate the technique of cartesian coordinates, which, for the most part, has been absorbed on the one hand in the notion ("set-theoretic" or "categorical") of "product" and on the other hand in linear and multilinear algebra which has supplanted it, to great advantage, in the curriculum of the first years at the university.

end of this volume to the most accessible works for readers who desire to fix precisely their information on these ideas and, eventually, to understand the corresponding proofs.

I must mention here the help J. P. Serre has given me in drawing up this volume. His remarks, criticisms, and suggestions have been extremely helpful. It is rare, indeed, for the historian to have the cooperation of one of the principal actors in the events he relates, and I am deeply grateful to him.

I

Introduction

To facilitate the exposition, the history of algebraic geometry has been divided into seven "epochs":

ca. 400 B.C.–1630 A.D. *Prehistory.*
1630–1795 *Exploration.*
1795–1850 *The golden age of projective geometry.*
1850–1866 *Riemann and birational geometry.*
1866–1920 *Development and chaos.*
1920–1950 *New structures in algebraic geometry.*
1950– *Sheaves and schemes.*

A last chapter describing some recent results and some of the principal problems that are unsolved at this moment has been added.

Of course, this division has been introduced only for convenience and has no intrinsic value. Moreover, the reader will verify that the ideas and theories belonging to one epoch often have roots much farther down that extend almost invariably upwards.

Also, it is clear that there can be no question of undertaking, within the limits of this work, a truly detailed and comprehensive history of algebraic geometry; such an enterprise, which is certainly desirable to see realized some day, would be a work of huge dimensions. Thus, the author is satisfied to mention the ideas and facts that have seemed most important to him, and from this, it very evidently follows that our "historical overview" can have no pretense of absolute objectivity (if history can ever attain this ideal). The works cited in the bibliography will allow the reader to form a personal opinion on the points that seem doubtful or insufficiently supported by the facts.

The internal references are given in the form (A, n), where A is the number of the chapter in roman numerals and n is the number of the section in arabic numerals.

Table of Notations

$\mathbf{B}(D)$	vector bundle associated to a divisor D: VIII, 4.
$\mathbf{C}$	field of complex numbers.
$C'.C''$	"intersection cycle" of two algebraic cycles on a variety: VI, 37 and VII, 28.
$(C'.C'')$	"intersection number" of two algebraic cycles of complementary dimensions: VI, 37 and VII, 28.
$c_j(\mathbf{E}), c(\mathbf{E})$	Chern class of index j, total Chern class of a vector bundle $\mathbf{E}$: VIII, 5.
c_j	Chern class of index j of a holomorphic manifold: VIII, 5.
$\chi(D)$	Euler–Poincaré characteristic of a divisor: VIII, 12.
$\|D\|$	complete linear system generated by a divisor: VI, 9 and VII, 43.
$D + D', -D$	operations on divisors: VI, 6 and VI, 43.
$D \leqslant D'$	order relation on divisors: VI, 6 and VI, 43.
$\deg(D)$	degree of a divisor on a curve: VI, 6.
dz	differential of an element $z \in K$ (K a finite extension of $\mathbf{C}(X)$): VI, 12.
$d\omega$	exterior differential of a p-form: VII, 2.
$\Delta(\omega), \Delta$	divisor of a meromorphic differential form of maximal degree (canonical divisor): VI, 13 and VI, 44.
∂	boundary of a chain: VI, 36.
$\underset{a}{\sim}, \underset{h}{\sim}, \underset{n}{\sim}$	algebraic equivalence, homological equivalence, numerical equivalence: VI, 36; VII, 21; VII, 53.
$(f)_0, (f)_\infty, (f)$	divisor of zeros, divisor of poles, principal divisor of a function $f \in K$: VI, 8 and VII, 43.
$\mathscr{F}(U)$	set of sections of a sheaf over U: VIII, 8.
$\mathscr{F}_x$	fiber of a sheaf at a point: VIII, 8.
$\mathscr{F} + \mathscr{G}, \mathscr{F} \otimes \mathscr{G}, \check{\mathscr{F}}, f\mathscr{F}, \mathscr{F}/\mathscr{G}$	operations on sheaves: VIII, 8.
$\mathbf{F}_q$	finite field with q elements.
$\Gamma(U, \mathbf{E})$	set of sections of a vector bundle over an open set U: VIII, 2.

$\Gamma(\mathbf{E})$	set of sections of a vector bundle over the total space: VIII, 2.
$\mathbf{G}, \mathbf{G}_a, \mathbf{G}_n, \mathbf{G}_l$	groups of divisors: VII, 52.

$H_j(M, \mathbf{Z})$, $H_j(M, \mathbf{Q})$, $H_j(M, \mathbf{R})$, $H_j(M, \mathbf{C})$ homology groups and spaces: VI, 36.

$H_\bullet(M, \mathbf{Z})$, $H_\bullet(M, \mathbf{Q})$, $H_\bullet(M, \mathbf{R})$, $H_\bullet(M, \mathbf{C})$ homology rings: VI, 36.

$H^j(M, \mathbf{Z})$, $H^j(M, \mathbf{Q})$, $H^j(M, \mathbf{R})$, $H^j(M, \mathbf{C})$ cohomology groups and spaces: VII, 3.

$H^\bullet(M, \mathbf{Z})$, $H^\bullet(M, \mathbf{Q})$, $H^\bullet(M, \mathbf{R})$, $H^\bullet(M, \mathbf{C})$ cohomology rings: VII, 3.

$H^0(U, \mathscr{F})$	set of sections of a sheaf over an open set U: VIII, 8.
$H^j(\mathscr{F})$, $H^j(X, \mathscr{F})$	cohomology groups of a space X with values in the sheaf $\mathscr{F}$: VIII, 10.
$\mathbf{H}^p$, $\mathbf{H}^{r,s}$	space of harmonic forms: VII, 4 and VII, 7.
$h^{r,s}$	dimension of $\mathbf{H}^{r,s}$: VII, 7.
$\int_C \omega$, $\langle C, \omega \rangle$	integral of a form over a chain: VII, 2.
L(D)	space of $f \in K$ such that $(f) + D \geqslant 0$: VI, 6 and VI, 43.
$\mathbf{N}$	set of integers $\geqslant 0$.
$\mathscr{O}_M$, $\mathscr{O}_x$	structure sheaf of a variety, of a scheme, fiber of this sheaf at a point: VIII, 7; VIII, 8; VIII, 27; VIII, 28.
$\mathscr{O}(\mathbf{E})$	sheaf of germs of sections of a vector bundle: VIII, 7.
$\mathscr{O}_M(D)$	sheaf associated to a divisor: VIII, 8 and IX, 60.
Ω^p_M	sheaf of germs of p-forms: VIII, 7.
$\mathbf{P}(\mathscr{E})$	projective scheme associated to a quasi-coherent $\mathscr{O}_S$ − Module $\mathscr{E}$: VIII, 43.
$\mathbf{P}_N(\mathbf{C})$	complex projective space of dimension N.
$\mathbf{P}_N(k)$	projective space of dimension N over a field k: VII, 14.
p_g	geometric genus: VI, 46.
p_a	arithmetic genus: VI, 46; VII, 17; VIII, 6; VIII, 10.
$p^{(1)}$	linear genus: VI, 47 and VIII, 16.
P_k	plurigenus: VI, 47 and VIII, 16.
$\mathbf{Q}$	field of rational numbers.
q	irregularity: VI, 48 and VIII, 17 and 18.
$\mathbf{R}$	field of real numbers.
R_j	jth Betti number: VI, 36.
$\mathbf{Z}$	set of positive and negative integers.

II — THE FIRST EPOCH

Prehistory
(ca. 400 B.C.–1630 A.D.)

1. It is difficult to speak of the history of algebraic geometry before the essential objects of this science, algebraic curves and surfaces at first and later general algebraic varieties, had been clearly conceived; such was the case up until the discovery of "cartesian coordinates" by Descartes and Fermat. Nevertheless, in several respects, the development of geometry by the Greeks prepared the way for the study of algebraic varieties.

2. In the first place, contrary to what is often believed, the geometry of the Greeks is quite the reverse of "pure" geometry, that is, geometry from which the calculations have been all but eliminated; such an idea appears first in the nineteenth century (see (IV, 4)). In fact, the Greeks cannot even separate algebra from geometry since algebra for them is essentially "geometric": it must be remembered that they do not calculate with numbers but with magnitudes and their relations, and that when they multiply two lengths, they obtain a magnitude of another type, namely, an area. Of course, this point of view together with the absence of convenient algebraic notation condemns this algebra to remain in an embryonic state. The general notion of polynomial is inconceivable in the framework of Greek mathematics.

But, on the other hand, this conception of algebra leads the Greeks to constantly mix arguments we consider "geometric" with calculations very analogous to those of our cartesian coordinates. In fact, it is sometimes by reference to two coordinate axes that they study the properties of a curve; this is the procedure Apollonius uses most often for conics in taking a tangent and the corresponding diameter as coordinate axes so that the foci of the conic enter only in an auxiliary way.

3. Moreover, the Greeks' favorite method of resolution of algebraic problems is to obtain a solution by the intersection of auxiliary curves. A typical example is the solution of the famous problem of *doubling the cube* by the method that led to the discovery of the conics. The problem is to find a length x such that $x^3/a^3 = b/a$ where a and b are two fixed lengths; Hippocrates of Chios (about 420 B.C.) changed the

problem into a "double proportion" $a/x = x/y = y/b$ for two unknown lengths x, y. Menaechmus (about 350 B.C.), a student of Eudoxus, had the idea of referring the loci given by the equations $ay = x^2$ and $xy = ab$ to two coordinate axes, the coordinates x, y of the point of intersection of the two curves giving the solution to the problem. Only much later, did the same Menaechmus recognize that these curves could also be obtained as the intersection of a cone of revolution and a plane perpendicular to a generator of the cone.

4. In fact, the idea of geometric solutions of algebraic problems seems to date back to the last third of the fifth century B.C. It stimulated the Greek mathematicians to invent diverse curves, both algebraic and transcendental (without, of course, being able to make such a distinction between them): the quadratrix of Hippias (at the end of the fifth century B.C.) with polar equation $\rho = c(\phi/\sin \phi)$, which was used for the squaring of the circle and the trisection of an angle; the conchoid of Nicomedes (about 250 B.C.), a curve of fourth degree with polar equation $\rho = a + b/\cos \phi$, used for several problems that today we reduce to equations of the third degree; the cissoid of Diocles (about 150 B.C.), a cubic with a cusp, and polar equation $\rho = 2a \sin^2 \phi/\cos \phi$, which gave another solution of the doubling of the cube. The strength of this tradition can be measured when, in the middle of the seventeenth century, Descartes and Newton proclaim anew that the principal interest of algebraic curves is that they give geometric solutions to algebraic equations by means of the intersection of curves of the smallest degree possible.

5. Nevertheless, in the geometry of the Greeks, algebraic curves are also introduced as "loci" with respect to problems of purely geometric origin. The most beautiful example is undoubtedly the most profound part of Apollonius' work on conics, the study of normals to conics, in which the evolutes of conics are completely characterized and studied. Apollonius' theorems translate immediately in our notation into the equation of the evolute that only the undeveloped state of Greek algebra prevents him from writing. Moreover, it is known that the greatest part of the original work of the Greek mathematicians is lost and it is therefore difficult to get a clear idea of the extent of their knowledge. For example, several commentators allude to work on plane sections of a torus, curves of the fourth degree named "spiric sections of Perseus," but none of this work has survived.

6. There are more serious gaps in our information concerning surfaces and space curves. Apart from the plane, it seems that the only surfaces studied during the classical age are the quadrics of revolution and the torus; the latter appears even before the third century in Archytas' bold construction for doubling the cube by means of the intersection of a cylinder of revolution, a cone of revolution, and a torus. At a later time other surfaces such as helicoides and more general ruled surfaces as well as their intersections with other surfaces appear to have been considered (by whom is not known). But the only space curves precisely described

in the texts in our possession are, in addition to the circular helix, intersections of quadrics of revolution such as the "hippopede" of Eudoxus, which is the intersection of a sphere and of a cylinder of revolution that is tangent to it. The description of this curve as the locus of a point on a sphere moving with uniform rotation around a diameter which is itself rotating uniformly around a fixed axis can be considered as the first example (in geometric form) of a parametric representation of an algebraic space curve.

III — THE SECOND EPOCH

Exploration
(1630–1795)

1. This second period has a very precise origin, the invention (by Descartes and Fermat, independently) of cartesian coordinates. By giving a systematic procedure for translating each geometric relation between points in the plane or in space into a relation between coordinates of these points, the method of cartesian coordinates, aided by progress in algebraic notation and by the development of infinitesimal calculus, allows the simultaneous birth of algebraic geometry and differential geometry. In particular, with the clear understanding of the concept of a polynomial of arbitrary degree, Descartes can forthwith define the notion of algebraic curve that will easily generalize to surfaces. The idea of dimension, while clearly intuitive in this period, is however explicitly linked by Fermat to the fact that a plane curve is represented by a single equation $F(x, y) = 0$ and a surface in space by a single equation $F(x, y, z) = 0$. He seems even to catch a glimpse of the possibility that arbitrary dimensions might be defined in this way; in the eighteenth century, the first algebraic space curves will be defined as the intersection of two surfaces, therefore characterized by two equations $F_1(x, y, z) = 0$, $F_2(x, y, z) = 0$.

2. This period can be considered as dedicated to the exploration of the new territory opened by Descartes and Fermat in which the concern is as much the pursuit of general laws as the study of particular curves and surfaces. The first invariant notion to appear is *degree*: Descartes already knows that the degree of an algebraic plane curve is invariant under change of axes and Newton adds that it is equally invariant under central projection from one plane to another. By the eighteenth century, the invariance of the degree of a surface under change of axes and certain types of linear transformations is well known.

3. The general notion of *parametric representation* of a curve is at the very foundation of the ideas of Barrow and Newton on infinitesimal calculus. In the eighteenth century, Euler begins to avail himself of it in algebraic geometry and obtains, in certain cases, rational parametric representations of a curve from its cartesian equation. Of course, the condition for the existence of such a representation will not be found before Riemann.

4. Needless to say, the sharp distinction made today between algebraic geometry and differential geometry does not exist before the end of the nineteenth century and many of the notions pertaining to algebraic curves and surfaces are first introduced as particular cases of notions relating to differentiable curves or surfaces. Such is the case for the notion of tangent. The equation

$$(x - x_0)\,F'_x(x_0, y_0) + (y - y_0)\,F'_y(x_0, y_0) = 0$$

of the tangent at a point (x_0, y_0) to a plane curve with equation $F(x, y) = 0$ was known (apart from notation) from the middle of the seventeenth century. Fermat also defines points of inflection and gives a differential criterion to obtain them explicitly. (This criterion is equivalent, for a curve with equation $y = f(x)$, to the vanishing of the second derivative $f''(x)$ even though the general idea of derivative does not emerge until Newton and Leibniz.)

5. The study of tangents to a curve $F(x, y) = 0$ leads inevitably to the consideration of the *singular points* of the curve, where, simultaneously, $F'_x = 0$ and $F'_y = 0$. The fundamental idea that will dominate all further research on this subject is due to Newton, namely that in the neighborhood of such a point (x_0, y_0) the curve has several "branches" along which $y - y_0$ can be expanded in a series

(1) $$a(x - x_0)^{\alpha} + b(x - x_0)^{\alpha+\beta} + \cdots$$

with increasing exponents α, $\alpha + \beta, \ldots$, not necessarily integers. To him is due also the famous "polygon rule" by which the first exponent α is determined from the equation of the curve. Given the algebraic curve $F(x, y) = 0$ and $x_0 = y_0 = 0$, consider the monomials $Ax^m y^n$ appearing in $F(x, y)$ with nonzero coefficient. For each such term, take the corresponding point (m, n) in the plane and consider the convex broken line L, each side of which contains at least two of these points, the others all being on the same side of the line containing the side. Then, the numbers $-1/\alpha$ are the *slopes* of the sides of L. Newton seems to be unaware of the fact that *several* branches can correspond to the same exponent α; this point will be clarified by Cramer who also shows how Newton's method gives the higher exponents in the expansion (1). But time must pass before the various forms assumed by the diverse branches are understood; even the existence for algebraic curves of cusp points of the "second kind" is for a moment in doubt at the beginning of the eighteenth century. The study of "infinite branches" of curves is reduced to that of a point at a finite distance by a change of variables, most often $x' = 1/x$, $y' = y/x$, corresponding to the geometric notion of central projection.

6. We have seen (II, 4) the interest attached to intersections of algebraic curves in the seventeenth century; a problem that arises immediately is the evaluation of the number of points of intersection of two algebraic plane curves $P(x, y) = 0$, $Q(x, y) = 0$ of degrees m, n, respectively. To obtain the abscissas of these points, it is necessary "to eliminate" y from the two equations (considered as algebraic

equations in y with coefficients polynomials in x). Newton and Leibniz (and even Fermat before them) already possess general methods of elimination but it is solely Maclaurin, in the eighteenth century, who conjectures that the number of points of intersection must "in general" be mn (a known result of Apollonius for $m = n = 2$). Euler, who makes fruitless attempts to prove this conjecture, understands the difficulties caused by the presence of multiple or imaginary roots of the equation giving the abscissas of the points of intersection or by the reduction of the degree of this equation. He conceives the possibility of establishing the conjecture by the introduction of "imaginary points" or of "points of infinity," but he does not push this idea beyond consideration of a few particular cases. The first general result toward what we now call "Bezout's theorem" is effectively due to Bezout (around 1765), but it overlooks the problems raised by Euler. Using a new method of elimination, Bezout restricts to the case where the curves have no common asymptotic directions and proves that the equation of the abscissas is always of degree mn (certain vague points in his proof can be made entirely rigorous). Moreover, his method is presented for any number N of algebraic equations in N variables.

7. We will see later (IV, 13 and VII, 21) the importance in the algebraic geometry of the nineteenth century of the problems of "enumeration" linked to the theory of elimination, but it must be pointed out that in the middle of the eighteenth century the number of tangents to a plane curve of degree n issuing from a point (a, b) is known to be, in general, $n(n-1)$: indeed, their points of contact are the points of intersection of the given curve $F(x, y) = 0$ with the "polar" curve:

$$(a - x)F'_x(x, y) + (b - y)F'_y(x, y) + nF(x, y) = 0$$

which is of degree $n - 1$. For his part, Maclaurin remarks that an indecomposable curve of degree n cannot have more than $(n-1)(n-2)/2$ double points.

8. This is a suitable place to mention a problem associated to the question of algebraic plane curves that, in the middle of the eighteenth century, gives birth to the theory of determinants. Since the general equation $F(x, y) = 0$ of a curve of degree n has $n(n+3)/2$ arbitrary coefficients (F is determined only up to a scalar multiple so one of the coefficients can be fixed), there is in general only one curve of degree n passing through $n(n+3)/2$ points in the plane; however two arbitrary plane curves of degree n have "in general" $n^2 > n(n+3)/2$ points in common. This is "Cramer's paradox." Its resolution by Euler marked the beginning of the general theory of systems of linear equations.

9. In addition to these first general results, many more specific studies issue during this period to notably enrich the geometric material inherited from the Greeks. Fermat recognizes that the curves of second degree are none other than the conics and shows how to classify them by their equations. In the eighteenth century, Euler will classify the quadrics in the same way. In a celebrated work,

Newton attacks the classification of plane cubics and identifies 72 "types" mutually distinct under change of axes (6 other types that he had not found were added to this classification by his successors). No doubt inspired by the description of the conics as sections of a circular cone, Newton shows further how all these cubics are obtained by central projection from only 5 types. Euler undertakes an analogous classification of plane quartics, but the large number of possible cases leads him to abandon this task. (In the nineteenth century, Plücker will obtain 152 types of plane quartics mutually distinct under change of axes.)

10. Finally the study (inspired, undoubtedly, by that of the conics) of the geometric properties of certain algebraic curves of degree 3 or 4 is initiated. While Maclaurin obtains the first general theorems on cubics (for example, the fact that the line joining two points of inflection contains a third one), new curves such as the ovals of Cassini, the folium of Descartes, Bernoulli's lemniscate, Pascal's limaçon that are "remarkable" for their form or their metric or differential properties are added to those known to the Greeks.

IV — THE THIRD EPOCH

The Golden Age of Projective Geometry (1795–1850)

1. The beginning of this period is marked by a change in point of view that is rather abrupt by earlier standards. In less than 25 years, starting from the publication of Monge's *Géométrie* (1795) and owing to the work of Monge's school, especially of Poncelet, a tremendous expansion of all geometric concepts occurs with the systematic introduction of points at infinity and imaginary points. The movement is so successful that, for close to 100 years, "geometry" will mean the geometry of the complex projective plane $\mathbf{P}_2(\mathbf{C})$ or of complex projective three-space $\mathbf{P}_3(\mathbf{C})$.

2. We have already seen the sporadic appearance, from Newton to Euler, of the idea that a thorough study of an algebraic curve must take into account not only "points" in the usual sense but also "points at infinity" and "imaginary points" despite the lack of a precise definition for these "fictitious" points. But in fact, at least as far as *real* projective geometry is concerned, a rather complete "synthetic" theory of the projective plane had been devised by G. Desargues toward the middle of the seventeenth century. Seeking a mathematical foundation for the technique of "perspective" used by painters and architects, Desargues, by systematic use of central projection, formed a clear idea of the adjunction of a "line at infinity" to the usual plane so that his conception of $\mathbf{P}_2(\mathbf{R})$ is the same as the modern definition: the set of lines through the origin in $\mathbf{R}^3$. Desargues's originality is manifest above all in his use of central projection to pass from special cases to the general statement of a theorem, for example, from theorems on circles to general theorems on conics. It is in this way that Pascal, greatly influenced by Desargues's ideas, will prove his famous theorem on the hexagon. But Desargues's language, rather different from that of other mathematicians of his time, and the poor diffusion of his work (which was believed lost for some time in the nineteenth century) destined his ideas to be ignored for a long period.

3. The success of projective ideas, in the forefront since 1820, is due above all to the extensive simplifications that they allowed in permitting the statement of a theorem in all generality without the encumbrance of the exceptional cases. For example, although the intersection of two circles in the plane appears to be an

exception to "Bezout's theorem" for conics, the number of points of intersection becomes equal to 4 due to the fact that in $\mathbf{P}_2(\mathbf{C})$ all circles pass through two fixed imaginary points at infinity (the "cyclic points"). All nondegenerate conics in $\mathbf{P}_2(\mathbf{C})$ are projectively equivalent, as are all nondegenerate quadrics in $\mathbf{P}_3(\mathbf{C})$; the "types" of plane cubics are reduced to three: the nonsingular cubics, the cubics with a double point, and the cubics with a cusp point, etc. There results a renewed impetus to study all the problems in algebraic geometry that were sketched in the preceding epoch. In spite of a decisive turn toward "birational" geometry initiated by Riemann from 1850 (see (V, 14)), the great stream of projective geometry, properly speaking (in which, according to Klein, the only transformation group considered is the projective group), follows its course well beyond this time, up until the beginning of the twentieth century. It will elicit an enormous number of publications, much larger than all the other branches of mathematics, as can be verified by the number of pages devoted to it in the *Enzyklopädie* at the end of the century.

4. One of the attractions of complex projective geometry is its relative independence from algebra and the formal elegance that results, in contrast to the massiveness of most of the coordinate calculations of the preceding century. It is true that Möbius, Plücker, and Cayley give projective geometry a solid base by the use of homogeneous coordinates accompanied by a harmonious choice of indexing notation that maintains a symmetry and a clarity in the calculations so that they closely follow the geometric argument. But from the time of Poncelet, many geometers, notably Chasles in France and Steiner and von Staudt in Germany, systematically seek to eliminate all use of coordinates from their arguments. For example, the degree of a curve or surface will be obtained not by a computation but by a geometric argument that evaluates the number of points of intersection with an arbitrary line. Even the "metric" notions of euclidean geometry, lengths and angles, are admitted only when they have been translated into projective language in which the only numerical notion is the *cross ratio* of four collinear points, invariant under all projective transformations. This insistence on the "purity" of the methods frequently brings about remarkable results in a restricted realm such as the theory of foci and focal curves, to cite one example where the projective view exposes properties that appear accidental when seen only from the "metric" point of view. In addition, the tendency to restrict study to data invariant under what Klein will call the group of the geometry, prepares for the intervention of group theory that will become dominant at the end of the century. But the mirage of "purity" has drawbacks. It encourages the relaxation of the rigor of geometric argument in replacing (as Poncelet already does, fortunately, in instances of little importance) elementary algebraic arguments by more or less vague "principles" introduced without proof. In spite of the objections that Cauchy raises against the justification of such methods, they remained widespread for more than a century (see (VII, 19)). Moreover, without the aid of algebra, geometric intuition rapidly becomes powerless in the study of more com-

plex phenomena, especially in the approach to geometry in spaces of arbitrary dimension (see (IV, 9)). A typical example is the classification of linear transformations, which can hardly be accomplished without the theory of elementary divisors. It certainly seems that the methods of "pure geometry" have long delayed the recognition of the central role that linear and multilinear algebra should play in geometry and analysis.

5. At the most elementary level, projective geometry, from the very first, is concerned with the geometric properties of curves and surfaces of low degree. The classical theory of conics and quadrics is simplified and completed by a multitude of new properties concerning, above all, linear families ("pencils" and "webs" of conics and of quadrics) as well as "tangential" properties revealed by the principle of duality, one of the fundamental discoveries of the new school. With regard to quadrics and their intersections, the first systematic studies of space curves appear, beginning with the simplest, the twisted cubic, an example of the intersection of two surfaces having two irreducible components; next come two types of space quartics, the "biquadratic," the intersection of two quadrics, and the rational quartic, which is contained in only a single quadric.

6. The study of plane curves of the third or fourth degree is developed, modeled on the theory of conics and quadrics. "Cramer's paradox" applied to cubics shows, for example, that a cubic (irreducible or not), which passes through 8 of the points of intersection of two given cubics C, C′, necessarily passes through the ninth; in particular, if 6 of the intersection points are on a conic, the 3 others lie on a line. Pascal's theorem is the case where C and C′ are two systems of three lines. An analogous argument gives, for example, the properties of the configuration of the 9 points of inflection of a plane cubic lying 3 by 3 on 12 lines. Another theorem on cubics, which will be of more importance in birational geometry, is the theorem of Salmon on the invariance of the cross ratio of four tangents issuing from an arbitrary point on the curve. A similar study of plane quartics and surfaces of the third degree reveals the existence of still more remarkable "configurations" such as that of the 28 bitangents to a plane quartic or that of the 27 lines on a surface of third degree.

7. The common fate of most classifications, which is to become more and more complicated as soon as the number of parameters is increased, befalls projective geometry; the actual enumeration and the individual study of curves and surfaces of a given degree rapidly give way to the pursuit of general principles of classification, more theoretical than effective. The idea of duality introduces an entirely new projective invariant, the *class*, which is the number of tangents issuing from a point in the plane in "general position." An apparent new "paradox" arises here: a "general" curve of degree m has class $m' = m(m-1)$; by duality, the degree of the curve ought to be $m'(m'-1)$, which is absurd. The explanation is, of course, that the "general" curve of degree m is not "general" as a curve of class

$m' = m(m-1)$; it admits "tangential singularities," double tangents or points of inflection. The clarification of this situation results in the celebrated *Plücker formulas*: restricting to (irreducible) curves with "general" singularities (pointwise or tangential), that is, to curves with exactly d double points with distinct tangents, d' double tangents with distinct contact points, s cusp points of the first kind, and s' ordinary points of inflection. The degree m and the class m' of the curve are related to these numbers by the self-dual formulas:

$$m' = m(m-1) - 2d - 3s \tag{1}$$

$$m = m'(m'-1) - 2d' - 3s' \tag{2}$$

$$s' - s = 3(m' - m). \tag{3}$$

These formulas must be modified when the singularities are no longer "general," and it is only by means of the concept of "infinitely near singularities" (see (VI, 21)) that a general expression of formulas (1) and (2) linking the degree and the class will be found.

The extension of these ideas to space curves and to surfaces, by definition of systems of projectively invariant integers attached to these varieties, will be accomplished satisfactorily only at the end of nineteenth century in conjunction with the introduction of projective spaces of arbitrary dimension (see (VI, 24)).

8. With regard to plane algebraic curves, it is in the course of this same period that problems relating to *singular points* and to the *multiplicity of intersection* of two curves at a point (although being local questions, independent of the projective point of view) are definitively resolved due to progress in the theory of functions of a complex variable. In the development of his theory of analytic functions, Cauchy had met the problem of "ramification points" of a function where the analytic function under consideration does not return to its initial value, as the complex variable describes a little circle with center at such a point, but takes another "determination," the simplest being the function $z^{1/m}$ for m an integer $\geqslant 2$ at the ramification point $z = 0$. In combining this idea of "permutation of the determinations" with the polygon method of Newton (that up to that time had been applied only in the real domain), Puiseux, in 1850, obtains the description of an arbitrary singular point of a plane curve with equation:

$$F(x, y) = 0.$$

In the neighborhood of such a point (x_0, y_0), the curve decomposes into a finite number of "branches" each of which (after possible change of axes) admits a parametric representation of the form:

$$x - x_0 = t^p \tag{4}$$

$$y - y_0 = \sum_{\nu} c_\nu t^\nu \quad \text{(the series converges in the neighborhood of } t = 0) \tag{5}$$

where the exponents ν such that $c_\nu \neq 0$ are all $> p$ and are relatively prime.

At this time it would have been easy to deduce the definition of the multiplicity of intersection of two plane curves $F(x, y) = 0$, $G(x, y) = 0$ at a point (x_0, y_0) from Puiseux's theorem. In fact, it is only after Riemann's work and at the time when "linear systems of plane curves" (see (VI, 50)) are introduced that several geometers (Cayley, M. Noether, Halphen, etc.) consider this question. Writing the "branch" of $F(x, y) = 0$ defined by (4) and (5) in the "irrational" form:

(6) $$y - y_0 = \sum_{\nu} c_\nu (x - x_0)^{\nu/p} = \phi(x)$$

and the same for a branch of the curve $G(x, y) = 0$:

(7) $$y - y_0 = \sum_{\nu} c'_\nu (x - x_0)^{\nu/p'} = \psi(x)$$

the multiplicity of intersection of the two branches at, for simplicity, the point where $x_0 = 0$, is defined by forming the product

$$\prod_{i,j} (\phi(\omega^i x) - \psi(\omega'^j x))$$

where ω (resp. ω') is a primitive pth (resp. p'th) root of unity and $0 \leqslant i \leqslant p - 1$, $0 \leqslant j \leqslant p' - 1$. The exponent of x in the first term of the expansion of this product is the multiplicity sought. (More geometrically, the two branches are cut by a line parallel to Oy with the abscissa x tending toward 0; the multiplicity of intersection is the order, with respect to x, of the product of the pp' segments with endpoints a point of intersection with the branch of $F(x, y) = 0$ under consideration and a point of intersection with the branch of $G(x, y) = 0$ under consideration.) Then the total multiplicity is the sum of these multiplicities over all the branches of the two curves. For example, for the two lemniscates of Bernoulli:

$$(x^2 + y^2)^2 - (x^2 - y^2) = 0, \qquad (x^2 + y^2)^2 - (y^2 - x^2) = 0,$$

the first with two branches $y = x - 2x^3 + \cdots$, $y = -x + 2x^3 + \cdots$, and the second with two branches $y = x + 2x^3 + \cdots$, $y = -x - 2x^3 + \cdots$, the total multiplicity is 8 for the intersection at the origin. With this definition, Bezout's theorem for plane curves takes its definitive form: the sum of the intersection multiplicities for all the common points of the two curves (assumed irreducible and distinct) is the product of the degrees.

9. In the abundant harvest of results in complex projective geometry, prominence should be given to the *geometry of lines* and its generalizations as well as to the introduction of *n-dimensional projective geometry*, two currents of ideas that very rapidly, in the last third of the nineteenth century, blend into one.

The geometry of lines begins around 1830 with the study of "complexes" and "congruences" of lines in projective space that are, respectively, families of lines dependent on 3 and 2 parameters; the families dependent on only one parameter, namely, *ruled surfaces*, having been encountered much earlier, of course. In the study of these families, lines are considered no longer as sets of points but as objects dependent on 4 parameters. When Cayley and Plücker introduce, for a line defined by two distinct points with homogeneous coordinates:

$$\begin{pmatrix} x_0 & x_1 & x_2 & x_3 \\ y_0 & y_1 & y_2 & y_3 \end{pmatrix}, \tag{8}$$

the line "coordinates" $p_{ij}(0 \leqslant i,j \leqslant 3;\ i \neq j,\ p_{ij} = -p_{ij})$, which are the determinants of order 2 extracted from the matrix (8) and that satisfy the relation:

$$p_{01}p_{23} + p_{02}p_{31} + p_{03}p_{12} = 0, \tag{9}$$

the analogy between the set of lines and the "quadric" in the projective space $\mathbf{P}_5(\mathbf{C})$ defined by the homogeneous equation (9) is manifest. F. Klein makes the leap by showing how the theory of "complexes" and "congruences" is explained in terms of "geometry on the quadric." But, in fact, the introduction of the geometric language of the "space of n dimensions" in algebra had been formulated in 1845 by Grassmann and Cayley, and Grassmann's "exterior algebra" already contains, in substance, the definition of the "grassmannian," the set of linear projective varieties of dimension $r - 1$ in $\mathbf{P}_{n-1}(\mathbf{C})$, and its interpretation as a subvariety of a $\mathbf{P}_\mathrm{N}(\mathbf{C})$ of large dimension. (The quadric (9) is the particular case corresponding to $n = 4$ and $r = 2$.) Nevertheless, the comprehensive study of varieties contained in arbitrary projective spaces $\mathbf{P}_n(\mathbf{C})$ really begins around 1885 with C. Segre, who considers particularly quadrics and their subvarieties in these spaces; it is also he who is the first to introduce the concept of *product* of two varieties and to show that a product $\mathbf{P}_m(\mathbf{C}) \times \mathbf{P}_n(\mathbf{C})$ can be considered as a subvariety of a $\mathbf{P}_\mathrm{N}(\mathbf{C})$ by means of a "Segre map." About this same time, Schubert is motivated by problems in "enumerative geometry" (see (VII, 21)) to investigate in detail subvarieties of grassmannians, which will, in our time, become more important because of their role in algebraic topology.

The concept of algebraic variety of arbitrary dimension also allows objects other than lines or linear varieties to be considered as "points" in a projective space $\mathbf{P}_\mathrm{N}(\mathbf{C})$. For example, the set of hypersurfaces (irreducible or not) of fixed degree m in a projective space $\mathbf{P}_n(\mathbf{C})$ can be identified to the set of homogeneous polynomials, not identically zero, in $n + 1$ variables having degree m, with proportional polynomials considered the same: it is evidently the set $\mathbf{P}_\mathrm{N}(\mathbf{C})$ with $\mathrm{N} = \binom{n+m}{n} - 1$. It is more difficult to think of the set of space curves of fixed degree in three-space, for example, as an algebraic variety, but already in 1860 Cayley has the idea of associating to such a curve the "complex" of lines that meet it, a variety of codimension 1 in the space of lines. This is the first step toward what will become in the modern era, "Chow coordinates" (VII, 31).

10. Another essential contribution of the school of projective geometry is the fundamental role played by the idea of *transformation*. We have seen that, already in the seventeenth and eighteenth centuries, certain transformations had been used sporadically, but they appeared as hardly more than artifices of the calculation. In contrast, from the beginning of Poncelet's work, the emphasis is put on the search for properties invariant under the group of projective transformations, and

on their use to reduce proofs to special cases of particularly simple figures or to describe certain curves or surfaces conveniently. For example, a conic is described as the intersection of two variable lines passing through two fixed points, one line the image of the other under a fixed projective transformation. For a long time the only projective invariant that occurs is the cross ratio of 4 collinear points (or of 4 concurrent lines in a plane), but from about 1845 the general theory of *invariants and covariants* comes into existence. Its remarkable success in the years 1860–1870 appears momentarily to give an answer (at least in theory) to all the problems of projective algebraic geometry. For example, recall that the set of plane curves (irreducible or not) of fixed degree r can be identified with the projective space $\mathbf{P}_N(\mathbf{C})$, where $N = r(r+3)/2$ is the number, reduced by one, of coefficients of a homogeneous polynomial of degree r in 3 variables. The linear group $\mathbf{GL}(3, \mathbf{C})$ operates naturally on this space, and two plane curves of degree r are projectively equivalent if they belong to the same orbit under the group action. This makes precise the vague idea of curves of the same "type." The theory of invariants permits the explicit determination (by a very long but purely mechanical computation) of all the rational functions of the N parameters of the curve that are invariant under the action of $\mathbf{GL}(3, \mathbf{C})$: they are rational functions of a *finite* number of them (Hilbert's finiteness theorem). If the values of these invariants characterized an orbit, the problem of projective classification of plane curves of fixed degree would, in principle, be resolved; in fact, as Hilbert will show in 1893, certain types of curves escape this classification because *all* the invariants take the value zero there (*null forms*). For example, for $r = 5$, the curves having a quadruple point, or a triple point with a single tangent, or those that decompose into a quartic with a double point and a line tangent to this quartic at the double point cannot be distinguished from one another by the values of the invariants.

After a very active period, the theory of invariants will become dormant at the end of the nineteenth century due partly to the extreme complexity of the computations and partly to Hilbert's solution of the problems of finiteness, which had motivated most of the investigations. It is only quite recently that this theory has recovered its importance in algebraic geometry (IX, 14). It must be mentioned, however, that it is in the course of his work on the theory of invariants that Hilbert proved the first fundamental theorems in commutative algebra (VII, 16), thus preparing the way for all the modern applications.

11. From the years 1820–1830, the first nonlinear *birational transformations*, which take on their full importance only after Riemann, emerge to join the projective transformations. It is, above all, the *quadratic transformation*, in various forms (particularly under its metric form, the "inversion"), that is studied. In its simplest form, it maps the point with homogeneous coordinates x, y, z to the point with homogeneous coordinates x', y', z' determined (up to scalar multiple) by the relations:

$$\frac{x'}{yz} = \frac{y'}{zx} = \frac{z'}{xy}. \tag{10}$$

This transformation is bijective on the complement of the union of the three lines joining the points with homogeneous coordinates $(1, 0, 0)$, $(0, 1, 0)$ and $(0, 0, 1)$, but to each of the points on one of these lines, other than these three "base points," there corresponds the opposite "base point," and the point corresponding to a base point is evidently not defined. However, when two points tend toward a "base point" along two curves having distinct tangents, their transforms tend toward two distinct points on the line opposite the base point. The transformation is said to blow up the base point into the opposite line, a property that will become central for the problem of "resolution of singularities" (see (VI, 20), (VII, 44), (IX, 46)).

Special cases of birational transformations "from curve to curve" are also known at that time. An example is the transformation mapping a point on a curve to the point of contact of the tangent to the curve that corresponds to it by duality or (in euclidean geometry) to its center of curvature or (if the curve is rational) to the parameter of the point in a rational parametric representation, etc. But, in fact, no comprehensive study of this type of transformation emerges before 1850, with the exception of the case where the two curves are identical and birationally equivalent to the projective line, which is clearly the same notion as before for the plane, in one less dimension.

12. Nevertheless, rather quickly, this case is seen to be related to another type of very promising future, the idea of "multiple-valued" *correspondence* (traces of which are already found in Maclaurin and Euler). Generally speaking, for two strictly positive integers m, n an "(n, m) correspondence" on the projective line is an algebraic relation:

$$\mathrm{F}(x_0, x_1, y_0, y_1) = 0 \tag{11}$$

between the homogeneous coordinates (x_0, x_1) of a point x and the homogeneous coordinates (y_0, y_1) of a point y, where F is a polynomial homogeneous of degree m in x_0, x_1 and homogeneous of degree n in y_0, y_1; in addition, F is assumed irreducible, which implies, in particular, that there is no point x (resp. y) for which the relation (11) becomes indeterminate in y (resp. x). To each $x \in \mathbf{P}_1(\mathbf{C})$ correspond, by (11), n not necessarily distinct points $y \in \mathbf{P}_1(\mathbf{C})$, and to each $y \in \mathbf{P}_1(\mathbf{C})$ correspond, by (11), m not necessarily distinct points $x \in \mathbf{P}_1(\mathbf{C})$. Chasles's "principle of correspondence" says, then, that the number of points x (distinct or not) such that at least one of the corresponding points y is equal to x (the so-called "united points" of the correspondence) is *equal to* $m + n$, unless for *all* $x \in \mathbf{P}_1(\mathbf{C})$, at least one of the y is always equal to x; the principle results immediately from the substitution $y_0 = x_0, y_1 = x_1$ in (11).

A beautiful example of the application of this principle is given by Poncelet's "closure theorems." (Recall that a conic is a rational curve, the transform of $\mathbf{P}_1(\mathbf{C})$ by a bijective birational transformation.) Given two conics C, C′ in general position in the plane and an integer $n \geqslant 3$, define on C a (2, 2) correspondence in the following way: to a point M on C corresponds the nth point M_n of a sequence $\mathrm{M}_0 = \mathrm{M}, \mathrm{M}_1, \ldots, \mathrm{M}_n$ of points on C where each line $\mathrm{M}_i\mathrm{M}_{i+1}$ is tangent to C′. It is easy to verify that if n is even and $M_{n/2}$ is a common point of C and C′, then $M_n = M$; if, on the contrary, n is odd and $M_{(n-1)/2} = M_{(n+1)/2}$, and the tangent at this point is also a tangent to C′, then $M_n = M$. Thus, in all cases, there are at least 4 points M on C for which $M = M_n$. The principle of correspondence states that if there is a *single* other point M such that $M = M_n$, then for *all* points M on C, $M = M_n$ (in other words, if there is a polygon of n distinct sides inscribed in C and circumscribed about C′, there are, necessarily, infinitely many).

13. Most of the applications Chasles makes of his principle concern the determination, by the geometric procedure mentioned above, of the degree of curves given as "loci" of points in the plane subject to algebraic conditions. He determines the number of points of the curve on an arbitrary line by finding these points as "united points" under a (n, m) correspondence on the line. As his principal intereest is in the theory of conics, he associates to each family $\sum$ of conics, depending on a parameter, what he calls its "characteristics" μ, ν that are, respectively, the number of conics in the family passing through an arbitrary point in the plane and the number tangent to an arbitrary line in the plane. Thus, reasoning semi-empirically, he concludes that the degree of a locus of points depending on the variable conic in $\sum$ is most often of the form $a\mu + b\nu$, where a and b are constants depending on the locus under consideration. For example, let C be a fixed conic and consider the locus Γ of a point M whose polar with respect to C is also the polar with respect to a conic (depending on M) in the family $\sum$. To find the degree of Γ, Chasles looks for the number of points M of Γ on a line D. He considers the correspondence that to a point P of D associates the points of D that are poles (with respect to a conic in $\sum$) of the polar Δ of P with respect to C. It is easy to see that this is a (ν, μ) correspondence on D and consequently that the desired degree of the locus Γ is $\mu + \nu$.

With results of this nature, Chasles and de Jonquières inaugurate a branch of projective geometry called "enumerative geometry." It will develop considerably in the last third of the nineteenth century, and we will return to it later with regard to questions about the notions of multiplicity of intersection and of specialization when we discuss how to justify their use by the geometers of this school (VII, 21). Enumerative geometry deals with problems where "in general" the number of solutions is finite and constant, and it is necessary to determine this number under conditions inaccessible by the usual algebraic calculations. For example, Chasles considers the problem of determining the conics tangent to 5 fixed conics in "general position" and shows that there are 3,264 of them. It is interesting to sketch his ingenious argument. With the notation introduced above, the points of contact of the conics of $\sum$ tangent to C are exactly those where Γ meets C, therefore there are in general $2(\mu + \nu)$ of them. Applying this first to the family $\sum'$ of conics passing through 4 given points in general position, the characteristics of $\sum'$ are easily found to be $\mu = 1$, $\nu = 2$ so that $2(\mu + \nu) = 6$; and then to the family $\sum''$ of conics passing through 3 given points and tangent to a given line, the "characteristics" of $\sum''$ are found to be $\mu = 2$, $\nu = 4$ so $2(\mu + \nu) = 12$. If $\sum_1$ is the family of conics passing through 3 points and tangent to C, its "characteristics" are, therefore, (6, 12). Proceeding in this way, if $\sum_2$ is the family of conics passing through 2 given points and tangent to 2 given conics, its characteristics are (36, 56); if $\sum_3$ is the family of conics passing through 1 given point and tangent to 3 given conics, its characteristics are (184, 224); lastly, if $\sum_4$ is the family of conics tangent to 4 given conics, its characteristics are (816, 816); hence, finally, the desired number $2(816 + 816) = 3{,}264$.

V — THE FOURTH EPOCH

Riemann and Birational Geometry (1850–1866)

1. The period we reach now is without any doubt the most important of all in the history of algebraic geometry to this day. It is entirely stamped by the work of one man, Bernhard Riemann, one of the greatest mathematicians who ever lived, and also one of those who have had, most profoundly, the perception (or divination) of the essential *unity* of mathematics. It is quite a paradox that in the work of this prodigious genius, out of which algebraic geometry emerges entirely regenerated, there is almost no mention of algebraic curve; it is from his theory of algebraic *functions* and their integrals that all of birational geometry of the nineteenth and the beginning of the twentieth century issues. We will see that, without Riemann himself having been clearly conscious of it, two of his other celebrated memoirs, that on the *Hypotheses of geometry* (i.e., "riemannian spaces") and that on the zeta function, would open new horizons to modern algebraic geometry from 1920 onward.

2. The origin of the theory of "abelian integrals" is the study of integrals of the form:

$$(1) \qquad u = \int_a^x \frac{R(t)\,dt}{\sqrt{P(t)}}$$

where $P(t)$ is a polynomial of degree 3 or 4 and $R(t)$ is a rational function. One such integral expresses the length of an arc of the ellipse (whence the name "elliptic integral"), another, the length of an arc of the lemniscate. At the beginning of the eighteenth century, it was maintained (without rigorous proof, as was the style of the time) that such integrals could not be expressed by algebraic or logarithmic functions. However, the idea persisted that there was some formula analogous to the one expressing the sum of the arcs of the circle:

$$\alpha + \beta = \arcsin(x\sqrt{1-y^2} + y\sqrt{1-x^2}),$$

$(x = \sin\alpha, y = \sin\beta)$, when the circle is replaced by an ellipse. It is exactly a formula of this type for arcs of the lemniscate that Fagnano discovered in 1714. In 1760, Euler, acquainted with Fagnano's work, showed that, in general, for an integral of the type (1), there is a relation of the form:

$$(2) \qquad \int_a^x \frac{R(t)\,dt}{\sqrt{P(t)}} + \int_a^y \frac{R(t)\,dt}{\sqrt{P(t)}} = \int_a^z \frac{R(t)\,dt}{\sqrt{P(t)}} + W(x,y)$$

where z is an *algebraic* function of x, y and W is either a rational function of x and y or the sum of a rational function and a logarithmic function of x and y, that is, a function of the form $\log S(x, y)$, where S is rational.

3. Abel's research on this question, which occurs at the same time as his discovery of elliptic functions around 1825, is directly connected to Euler's memoir. Whereas Euler had remarked on the impossibility of generalizing relation (2), while adhering to its form, when P is of degree $\geqslant 5$, Abel places himself, from the start, in a much more general framework, which allows him to get out of this apparent difficulty. In place of $\sqrt{P(x)}$, he considers, more generally, an algebraic function y of x, defined as an "implicit" function by an arbitrary polynomial equation $F(x, y) = 0$, and he replaces the integral (1) by what has been called, since then, an "abelian integral":

$$\Psi(x) = \int_a^x R(t, y)\, dt \tag{3}$$

where R is a rational function of two variables, and, in the integration, y must be replaced by the "implicit function" of t deduced from the equation $F(t, y) = 0$. Finally, the sum on the left side of (2) is replaced by a sum $\Psi(x_1) + \Psi(x_2) + \cdots + \Psi(x_m)$ of values of the integral at an arbitrary finite number of points. Of course, Abel does not hesitate to give imaginary values to the x_j (since, in the case of elliptic integrals, it is in this manner that he conceives the idea of "inversion" of the integral by a function of a complex variable). Since, to each fixed value of t, there corresponds several roots of y from the equation $F(t, y) = 0$, it is necessary to specify, in the definition of $\Psi(x)$, which of these roots is to be substituted for y in $R(t, y)$. Although at this time Cauchy had hardly begun his work on integrals of functions of a complex variable, the context shows that Abel meant by $\Psi(x_j)$ the line integral, which we write now $\int_\Gamma R(t, y)\, dt$, where Γ is a path in $\mathbf{C}$ originating at a and ending at x_j, and where, for y, is to be substituted the continuous *function* of t on Γ, taking at a one of the values b satisfying $F(a, b) = 0$ and such that, of course, $F(t, y(t)) = 0$ for all points t on Γ. If y_j is the value taken by this function when t reaches the endpoint x_j, the number $\Psi(x_j)$ is abbreviated $\int_{(a,b)}^{(x_j, y_j)} R(x, y)\, dx$, and the expression that Abel considers is written thus:

$$V = \int_{(a,b)}^{(x_1, y_1)} R(x, y)\, dx + \cdots + \int_{(a,b)}^{(x_m, y_m)} R(x, y)\, dx. \tag{4}$$

Abel's new idea is not to take arbitrary (x_j, y_j) but rather to take, from the start, in geometric terms, *all* the points of intersection of the curve $F(x, y) = 0$ with a variable curve $G(x, y, a_1, \ldots, a_r) = 0$ depending rationally on parameters $a_1, \ldots, a_r$. The result he obtains is, then, the famous "Abel's theorem" according to which, under these conditions, V is the *sum of a rational function and a logarithmic function of the parameters* $a_1, \ldots, a_r$.

4. It is remarkable that the proof given by Abel of such a general theorem is hardly more than an exercise in the theory of the symmetric functions of the roots of a polynomial. But, not content with that, Abel sets out to determine, in particular, the conditions under which the expression V is in fact *constant*. At the cost of rather complicated computations, he obtains for $R(x, y)$, the expression $Q(x, y)/(\partial F/\partial y)$ (which will be central in Riemann's theory), where $Q(x, y)$ is a polynomial subject to certain linear conditions depending only on the curve $F(x, y) = 0$, so that it depends linearly on a certain number γ of parameters. Further, returning to Euler's initial problem, Abel observes that r arbitrary points (x_j, y_j) on the curve $F(x, y) = 0$ determine $a_1, \ldots, a_r$ and consequently the $m - r$ other points; therefore (since his conditions for V to be constant do not involve G) a sum of r integrals:

$$\int_{(a,b)}^{(x_1,y_1)} \frac{Q(x,y)\,dx}{\dfrac{\partial F}{\partial y}} + \cdots + \int_{(a,b)}^{(x_r,y_r)} \frac{Q(x,y)\,dx}{\dfrac{\partial F}{\partial y}}, \tag{5}$$

with limits of integration (x_j, y_j) *arbitrary points* on the curve, can be written as a sum of $m - r$ analogous integrals, where the limits of integration are algebraic functions of the (x_j, y_j). The generalization of Euler's result lies, then, in the fact, proved by Abel, that the number $m - r$ has a *minimum that depends only on the curve* $F(x, y) = 0$, a minimum that is, in general, equal to γ. For the example of ultraelliptic integrals where F is the polynomial $y^2 - P(x)$ with P of degree $2m$ or $2m - 1$, Abel obtains the value $\gamma = m - 1$ showing clearly why Euler's original statement could not be generalized. Moreover, it is evident that Abel is very close to the notion of *genus* that Riemann will introduce. In general, the minimum of $m - r$ and the number γ are, in fact, equal to the genus of the curve $F(x, y) = 0$. But the condition $V = \text{constant}$ imposed by Abel is, unfortunately, not sufficient in general to characterize integrals of the first kind, for certain integrals of the third kind can have this property, too. Also, it appears that Abel did not look for relations between the numbers γ or $m - r$ and the periods of these integrals (which he knew existed).

5. Abel's detailed memoir, presented to the Academy of Sciences of Paris in 1826, was not published until 1841; up to then, only the first form of Abel's theorem and its application to ultraelliptic integrals had been published in two short notes. It is interesting to observe here that these are the results that finally allowed Jacobi, after long, fruitless efforts, to discover the natural generalization of "inversion" of elliptic integrals to ultraelliptic integrals (1) with P a polynomial of degree 5 or 6. The "natural" idea was to consider in (1) x as a function of u, as Abel and Jacobi (and Gauss before them in unpublished work) had done for elliptic integrals; but the existence of four periods (in general, linearly independent over $\mathbf{Q}$) led Jacobi to the conclusion that x could not be a *meromorphic* function of u. Abel's theorem showed him, finally, the route to follow: it is necessary to consider (for P of degree 5 or 6) *two* equations:

$$(6)\qquad \int_a^x \frac{dt}{\sqrt{P(t)}} + \int_a^y \frac{dt}{\sqrt{P(t)}} = u$$

$$\int_a^x \frac{t\,dt}{\sqrt{P(t)}} + \int_a^y \frac{t\,dt}{\sqrt{P(t)}} = v$$

and to "invert" them in order to express the symmetric functions $x + y$ and xy as functions of u and v. Abel's theorem provides an "addition formula" for these functions from which it follows that they are meromorphic, quadruply periodic functions of the complex variables u, v.

6. Before Riemann, most of the research related to Abel's adds little when compared to the work of Abel himself with the exception of an enigmatic page from the letter-testament of Galois (1832). The latter states essential results of Riemann on the classification of abelian integrals into three types, the relations between the number of periods of an integral of the first kind and the number of linearly independent integrals of the first and second kind, and finally, the formula for the exchange of the parameter and the argument for integrals of the third kind. No trace of proof of these results has been found in Galois's papers and it is quite unlikely that Riemann knew of this text.

7. Riemann's work, like that of Galois or Dirichlet, is immediately striking in its adherence to the concepts that are studied there; secondary or superfluous ideas and, especially, all but absolutely indispensable calculations are eliminated whenever possible. Confident of his interior vision, Riemann does not hesitate to proceed when the mathematical techniques of his time do not furnish him with the means of a rigorous proof. He has been reproached for it, but all later developments have justified his intuitions once the necessary tools had been invented.

The works we consider here offer a dazzling succession of original ideas in which Riemann, almost systematically, "thinks off the beaten track" (following Hadamard's expression), approaching each problem in a way not envisaged by his predecessors.

8. His first work, the inaugural dissertation of 1851, is oriented (as he himself said) toward the study of algebraic functions and their integrals, but is placed, from the start, in a larger framework. In order to calculate multiple-valued functions of a complex variable (IV, 8), Cauchy and his students imagined "lines of obstruction" (Cauchy) or "cuts" (Hermite) traced in the plane so that, in the complement of their union, the diverse "determinations" of the function could not be interchanged. Against this artificial dissection of the plane in which diverse determinations tend to be regarded as distinct functions, Riemann puts forth, from the beginning, his conception of a connected surface with many sheets spread out over the plane in such a way that, above a point in the plane other than the ramification points, there are as many sheets as the function under consideration has "determinations;" the sheets join together above each of the ramification

points so that the determinations of the function give a *single* continuous function on the surface. To tell the truth, if the way he uses this concept shows that he had clear understanding of it, Riemann tries neither to describe it more precisely nor to show that such an object exists even for the case of the Riemann surface of an algebraic function $s(z)$ defined ("implicitly") by a polynomial equation $F(s, z) = 0$. For 50 years, his successors sought to fill this gap in trying, intuitively, to give the Riemann surface (most often restricting to the case of algebraic functions) an image in three-dimensional space in which Cauchy's "lines of obstruction" become lines where the sheets "cross themselves" without having points in common! In the case where the curve $F(s, z) = 0$ has no singularity, a natural image of the Riemann surface is, very simply, the subvariety of the space (of four real dimensions) $\mathbf{C}^2$ defined by $F(s, z) = 0$ (at least when points at infinity are disregarded); but, in the general case, Riemann imagines that above a ramification point there are as many points of the surface as *branches* in the sense of Puiseux, so that, for example, above a double point with distinct tangents there are two sheets without common point (in modern terms, Riemann considers the Riemann surface of the "desingularized" (or "normalized" (VII, 44)) curve). These difficulties will not be resolved until 1913 when H. Weyl, using the new tools of topology, will free the concept of Riemann surface from all recourse to an embedding in any space, numerical or projective, and will definitively establish the theory on a solid basis.

9. Moreover, Riemann's essential idea is that a Riemann surface must not be tied to a particular "multiple-valued" function but be considered *in abstracto*, as is the complex plane, as geometric-analytic substratum where analytic functions can be defined (on all or or on part of the surface); that is, it must be what we call, since the time of H. Weyl, a holomorphic manifold of (complex) dimension 1. Riemann's conception of analytic function differs from that of Cauchy or Weierstrass: instead of defining them by the existence at every point of a local expansion in power series, he prefers to consider them as complex functions $U + iV$ where the real functions U, V are related by the Cauchy-Riemann equations (the usual Cauchy conditions at points other than ramification points; at the latter, it is necessary first to express the function using the complex "uniformizing parameter" t that appears in the Puiseux expansions (IV, 8)). Moreover, Riemann is the first mathematician who exploits the fact that U and V are, then, harmonic functions.

10. In the first place, Riemann intends to extend Cauchy's theory, based on the study of line integrals $\int f(z)\,dz$ in the complex plane, to a Riemann surface S. Cauchy's theorem, which states that $\int_\Gamma f(z)\,dz$ along a closed path is zero for all holomorphic functions, is not valid in general, and Riemann discovers the reason for this is the fact that the surface S is not simply connected. He then is inspired to "make S simply connected" by removing a certain number of curves C_j (*Querschnitte*) from S so that the open complement $S' = S - \bigcup_j C_j$ becomes simply

connected. The integral of a function *holomorphic in S′* is again well defined. But a point of a curve C_j can be approached by points of S′ from one side or the other of C_j, and the limiting values of the integral are not, in general, the same from both sides. Still, if the integrand f is *meromorphic* in S and if all its poles are on the C_j, the local validity of Cauchy's theorem shows that the difference between these two limiting values is constant along all portions of C_j not containing a pole of the function; this difference is called a *period* of the integral. As for what happens when a point $z \in S'$ tends toward a pole of f, it is easy to see that $\int f(z)\,dz$ has, in the neighborhood of this point, a principal part that is, up to a constant factor, either a power t^{-k} of the uniformizer t (k an integer > 0) or a term $\log t$.

11. At this point, Riemann abruptly "reverses the engine," so to speak, and seeks, in a single stroke, to characterize *all* the analytic functions in S′ that exhibit, in the neighborhood of points of the C_j, the same behavior as the integrals $\int f(z)\,dz$. To develop his method easily, he restricts himself to an open, relatively compact subset T of S, bounded by a closed curve L and "made simply connected" following the same method with the *Querschnitte* C_j that, with L, determine the boundary of an open simply connected subset T′. To this data is added that of a certain number of points of T′, poles or logarithmic points of the functions sought, where the principal part of the functions is also given. Each point is joined to L or to one of the C_j by an arc so that, after removal of all these arcs, a simply connected domain T″ remains. Finally, Riemann takes (arbitrarily) a continuous *real* function ϕ on L and *real* constants a_j corresponding to each of the C_j. His fundamental theorem is that, up to an additive constant, there exists *one and only one function* $U + iV$ *analytic in* T″ having the given principal parts at the given poles or logarithmic points, and such that U reduces to ϕ on L and has, from each side of each C_j, limiting values differing by the "period" a_j. Being familiar with the integral methods of Gauss and Dirichlet in potential theory since the beginning of his studies at Göttingen, Riemann models his proof on Dirichlet's proof for the existence of harmonic functions in a three-dimensional domain taking given values on the boundary, by introducing a double integral dependent on an arbitrary function and by showing that the harmonic function U is the one giving the minimum of this integral. This is known as "Dirichlet's principle," the object of the most serious criticism by Riemann's contemporaries and justified only by Hilbert in 1900.

12. Armed with this extension of Cauchy's theory, Riemann can attack the problem of abelian integrals, the subject of his great memoir of 1857. He maintains the point of view of his dissertation, slightly modified to take points at infinity into account: this time the Riemann surfaces he considers are without boundary and are spread out not over the complex plane but over the "Riemann sphere" Σ obtained by compactifying **C** by the addition of a point at infinity (of course, the surface can have ramification points above the point at infinity of Σ). Here, the *Querschnitte* C_j are closed curves (which may be assumed to have a common point).

Riemann shows that the number of C_j necessary to make the surface simply connected is an even number $2p$; in modern terms, the C_j form a basis of the homology of S in dimension 1, and $2p$ is the "first Betti number." From the time of Clebsch, the invariant p has been called the *genus* of the Riemann surface.

As such a surface S is compact, the fundamental theorem in Riemann's dissertation can be applied without having to take the boundary L into consideration; therefore, the data consists only of the poles or logarithmic points with their principal parts and the real parts of the $2p$ periods corresponding to the C_j. From this, Riemann deduces his celebrated decomposition of "general" functions of the desired type into the sum of a constant and a linear combination of three fundamental types:

1° The (nonconstant) functions of the *first kind*, holomorphic in S′ and tending toward finite limits at the points of the C_j; they are linear combinations of p linearly independent ones.

2° The functions of the *second kind*, linear combinations of "elementary" functions, that have the same behavior at the points of the C_j as have functions of the first kind, but that have, in addition, a *single pole* in S′ of the first order with residue equal to 1.

3° The functions of the *third kind*, linear combinations of "elementary" functions, which here again have the usual behavior at points of the C_j, but have, moreover, *two* logarithmic points a, b in S′ with principal parts $\log(z-a)$ and $-\log(z-b)$.

Riemann treats functions having poles of order greater than the first as limits of functions of the second type having coincident poles of the first order.

13. All these developments may seem far removed from algebraic geometry; certainly, it is the manner in which Riemann rejoins algebraic geometry that has most astonished his contemporaries. Indeed, a function meromorphic in *the whole* of S is, clearly, a particular case of the preceding general functions; it is a function with no logarithmic point whose *periods are zero*. Such a function can be expressed, therefore, as a linear combination:

$$(7) \qquad s = c_1 W_1 + \cdots + c_p W_p + c_{p+1} E_{p+1} + \cdots + c_{p+r} E_{p+r} + c_{p+r+1}$$

where the c_j are constants, $(W_j)_{1 \leqslant j \leqslant p}$ is a basis for the vector space of functions of the first kind, and the E_j are an arbitrary number of "elementary" functions of the second kind, the c_j $(1 \leqslant j \leqslant p+r)$ being related by $2p$ homogeneous linear equations expressing the nullity of the periods. If S has n sheets, the symmetric functions of the values of one such function s at the n points of S above a point z of the Riemann sphere Σ are meromorphic functions of z on Σ, therefore *rational* functions of z; in other words, there is a polynomial relation $F(s, z) = 0$ identically satisfied on S, and the initial definition of the "algebraic functions" of Abel has been recovered. Moreover, if u is any one of the "multiple valued" functions of the three kinds considered by Riemann, the fact that the "jump" of such a function across a curve C_j is constant along C_j means that du/dz has no period and is, therefore, an algebraic function on S. In other words, the functions considered by Riemann are the "abelian integrals" in the original sense of Abel.

But the extraordinary decomposition (7) of the algebraic functions on S into a sum of transcendental "elementary functions" will, in Riemann's hands, extend to new notions central to all further developments.

In the first place, as the number of functions of the second kind in (7) is arbitrary, the r simple poles evidently can be chosen arbitrarily when $r \geqslant p + 1$. In particular, if Ω is an open connected subset of Σ, above which all of the sheets are without common point, a pole can be taken in each sheet so that these poles project onto distinct points of Ω. Then, for two sheets U_j, U_k above Ω, the set of points $z \in \Omega$ such that s takes the same value at the points of U_j and U_k projecting onto z cannot be contained in a nonempty open set. Otherwise, by the principle of analytic continuation, this set would be all of Ω, an absurdity due to the choice of the poles of s. Consequently, there is a nonempty open set $\Omega' \subset \Omega$ such that for $z \in \Omega'$, s takes n distinct values at the n points of S above z. The polynomial relation $F(s, z) = 0$ is, therefore, of degree n in s, and, moreover, F is *irreducible* since S is connected.

Now let u be a second meromorphic function on S. With the same notation, consider, for a $z \in \Omega'$, the n distinct values $s_1, \ldots, s_n$ taken by s at the points above z and the n corresponding values $u_1, \ldots, u_n$ of u at these points. It is clear that the functions:

$$u_1 + u_2 + \cdots + u_n = R_1(z)$$
$$u_1 s_1 + \cdots + u_n s_n = R_2(z)$$
$$\cdots\cdots\cdots\cdots\cdots\cdots\cdots\cdots$$
$$u_1 s_1^{n-1} + \cdots + u_n s_n^{n-1} = R_n(z)$$

are functions that continue analytically to meromorphic, hence rational, functions on Σ. From the fact that the Vandermonde determinant of the s_j is nonzero in Ω', by an elementary argument on symmetric functions, it follows that $u = R(s, z)$ where R is a *rational fraction*.

14. Thus, the abstract Riemann surface S is, in fact, identical to that of the algebraic function $s(z)$ defined by $F(s, z) = 0$, and Riemann attaches to it what will, after Dedekind's time, be called the *field of meromorphic* (or *rational*) *functions* on S. But, as his thought progresses, Riemann is inevitably led to consider two Riemann surfaces S, S_1 equivalent if there exists a bijective, biholomorphic correspondence between them. That amounts to considering two algebraic curves $F(s, z) = 0$, $F_1(s_1, z_1) = 0$ equivalent if there exists a birational correspondence between them; the purely algebraic formulation of this correspondence is that the fields of rational functions of S and S_1 are **C**-*isomorphic*. It is clear that the genus of S and that of S_1 are the same, and thus, is inaugurated *birational geometry* that will dominate all of algebraic geometry for the next 80 years: the emphasis here is put on properties that are preserved by birational transformations, and the projective invariants are superseded by the birational invariants.

15. Moreover, it is Riemann himself who, in the course of his memoir, poses and partially resolves, in a masterful way, the problems that will furnish the principal themes of this new geometry:

1° The expression of the rational functions on a curve C by means of the notion of "adjoint" curve: Riemann means by that an algebraic curve passing through the double points of the curve C, and expresses rational functions on C as quotients of the first (polynomial) members of the equations of two such curves. He shows, moreover, that the abelian integrals of the first kind are of the form

$$\int Q(s, z)\, dz/(\partial F/\partial s)$$

already introduced by Abel, but where $Q(s, z) = 0$ is the equation of an adjoint of degree $n - 2$ and $m - 2$ in s and z, respectively (if the degrees of the equation $F(s, z) = 0$ of C are n and m in s and z).

2° The determination of the rational functions having given poles and the corresponding "counting constants," which will be expressed in what we call the "theorem of Riemann-Roch" (VI, 14).

3° Counting parameters on which the isomorphism classes of Riemann surfaces of fixed genus depend (the "moduli problem") ((VI, 27) and (IX, 12 to 14)).

4° The definition of general theta functions and their use in the resolution of the problem of inversion of abelian integrals, which will give birth to the theory of abelian varieties (VII, 54).

We will examine in the following chapters Riemann's contributions to each of these questions as well as the developments made by his successors.

VI — THE FIFTH EPOCH

Development and Chaos (1866–1920)

In 1866, Riemann died prematurely at 40 years of age. While the extraordinary richness of his work in its new ideas and methods are to be a source of inspiration for a century, not one of his followers is able to proceed with the impressive synthesis he had outlined. Their tendencies and aptitudes attracted them to one aspect or another of Riemann's works, and thus are born several schools of algebraic geometry that tend to diverge up to the threshold of mutual incomprehension. At first, each of them expresses Riemann's results on the birational geometry of plane algebraic curves in its own language and attempts to give proofs of them in conformity with the principles of the school; then, with more or less success, it will seek to extend these methods to the study of algebraic surfaces and algebraic varieties of arbitrary dimension.

1. The algebraic school

1. Chronologically, this school comes last because it comes into existence only in 1882 with two fundamental memoirs, one by Kronecker, the other by Dedekind and Weber. But in the light of future history, it is the algebraic inclination that exercised the most profound influence on the birth of the concepts of modern algebraic geometry. In particular, as Riemann had exhibited the close connections between algebraic varieties and holomorphic manifolds, so from Kronecker and Dedekind-Weber dates the awareness of the profound analogies between algebraic geometry and the theory of algebraic numbers, which originated at the same time. Moreover, this conception of algebraic geometry is the most simple and most clear for us, trained as we are in the wielding of "abstract" algebraic notions: rings, ideals, modules, etc. But it is precisely this "abstract" character that repulsed most contemporaries, disconcerted as they were by not being able to recover the corresponding geometric notions easily. Thus the influence of the algebraic school remained very weak up until 1920.

2. Kronecker's work is destined, in the spirit of its author, to give solid algebraic foundation both to the theory of numbers and to algebraic geometry. It certainly seems that Kronecker was the first to dream of one vast algebraico-geometric

construction comprising these two theories at once; this dream has begun to be realized only recently, in our era, with the theory of schemes (see (VIII, 35)). In fact, Kronecker sketches only a few preliminaries in this ambitious design, centered essentially on the notions of *irreducible variety* and of *dimension*. It must be said that up until then, no one apparently thought it necessary to give precise definitions; geometric intuition seemed sufficient for geometry in a two- or three-dimensional space. For plane curves or surfaces in three-space, the decomposition of a curve (resp. surface) given by an arbitrary polynomial equation into irreducible curves (resp. surfaces) simply reflected the theorem (known essentially since Gauss) of the decomposition of a polynomial in $\mathbf{C}[T_1, \ldots, T_n]$ into irreducible factors. But we have seen (IV, 9) that it is in the last third of the nineteenth century that geometry in spaces of arbitrary dimension, where in the absence of "intuition" arguments founded on dimension were precarious, made its debut. Accordingly, Kronecker places the problem in its most general form: given a finite family of polynomials $P_\alpha \in \mathbf{C}[T_1, \ldots, T_n]$, the "algebraic variety" V, formed of the points $(x_1, \ldots, x_n)$ in $\mathbf{C}^n$ that annihilate all the P_α, must decompose into a finite number of "irreducible varieties" each having a well determined "dimension." To prove this, he proceeds by induction on n: it may be assumed that the P_α have no common factor and, by a change of coordinates, that each of the P_α has a term of (total) highest degree $m_\alpha = \deg(P_\alpha)$ of the form $c_\alpha T_1^{m_\alpha}$. By an ingenious method of elimination, Kronecker proves that the projection of V on $\mathbf{C}^{n-1}$ is an algebraic variety V′ defined in the same manner as V. The induction hypothesis and elementary arguments from the theory of algebraic extensions show that V is, indeed, a union of a finite number of algebraic varieties V_j $(1 \leqslant j \leqslant r)$ each of which can be defined by a "parametric representation":

$$x_i = R_i(y_1, \ldots, y_{d+1}) \qquad (1 \leqslant i \leqslant n),$$

where the $d+1$ variables y_j are related by an irreducible algebraic equation:

$$Q(y_1, \ldots, y_{d+1}) = 0,$$

and the R_i are rational functions. It is this number d that Kronecker calls the dimension (*Stufe*) of the irreducible component V_j. Moreover, he realizes that this procedure does not give intrinsic definitions and that everything must be expressed by means of the *ideal* $\mathfrak{a}$ (which is called a *Modulsystem* in his language) generated by the P_α in the ring $\mathbf{C}[T_1, \ldots, T_n]$. In particular, he sees already that if $\mathfrak{a}$ is prime, the variety V is irreducible (although, inadvertently, he takes an incorrect definition of prime ideal, which he will correct in a later memoir). The converse will be elucidated only by E. Lasker in 1905 with the introduction of the notion of primary ideal and the theorem on the decomposition of an ideal of $\mathbf{C}[T_1, \ldots, T_n]$ into an intersection of primary ideals. Thus, the criterion of irreducibility will be that if a product PQ of two polynomials is zero at every point of V, then at least one of them is zero at every point of V. It will be easier to obtain an intrinsic definition of dimension when V is irreducible, for then the rational functions on $\mathbf{C}^n$, which are defined at one point of V at least, have restrictions to V that form a *field* (the analogue of the field defined by Riemann for plane algebraic curves), and the dimension, in the sense of Kronecker, is simply the transcendence degree of this field over **C** (a notion that will be cleared up by Steinitz in 1910).

3. The Dedekind-Weber memoir has a much more restrained objective: restricting to curves, Dedekind and Weber propose to give purely algebraic proofs of all of Riemann's algebraic theorems. But their remarkable originality (which in all of the history of algebraic geometry is only scarcely surpassed by that of Riemann) leads them to introduce a series of ideas that will become fundamental in the modern era. Their starting point is the notion of field of rational functions that Riemann attaches to an isomorphism class of Riemann surfaces. Such a field K is a finite extension of the field **C**(X) of rational fractions over **C**, and the properties of a Riemann surface invariant under isomorphism must, therefore, be reflected precisely in the properties of K. Accordingly, Dedekind and Weber start with an abstract finite extension K of the field **C**(X) and propose to "reconstruct" a Riemann surface S such that K is isomorphic to the field of rational functions on S.

4. The progression of their ideas can be described as follows: *if* S were known, then a nonzero rational function $f \in K$ would have an order $v_{z_0}(f)$ at each point $z_0 \in S$, defined as the (positive or negative) exponent of the first nonzero term of the expansion $f(t) = \sum_\nu c_\nu t^\nu$ of f in series relative to the Puiseux "uniformizer" at the point z_0 (IV, 8). Then, for a *fixed* $z_0 \in S$, the mapping $f \mapsto v_{z_0}(f)$ is what we now call a *discrete valuation* on the field K. Recall that (for *any* field K) a discrete valuation is a *surjective* mapping $w : K^* \to \mathbf{Z}$ (K^* being the multiplicative group of nonzero elements of K) such that:

$$w(f+g) \geqslant \inf(w(f), w(g)) \qquad \text{if} \qquad f+g \neq 0$$

and

$$w(fg) = w(f) + w(g),$$

which imply:

$$w(1) = 0 \qquad \text{and} \qquad w(1/f) = -w(f).$$

By convention, w is extended to the entire field K by setting $w(0) = +\infty$; this avoids the restriction $f + g \neq 0$ in the first axiom. Then, aside from terminology, the crux of Dedekind and Weber's plan is to *define* the "points of the Riemann surface" to be the *discrete valuations* of the (abstractly given) field K.

To study these valuations, they note, first of all, that the restriction to the subfield **C**(X) of a valuation v of K is proportional to a discrete valuation w of **C**(X). But the discrete valuations of **C**(X) are easily determined. First, for all $\alpha \in \mathbf{C}^*$, necessarily $w(\alpha) = 0$ because, for each integer n, α has an nth root of unity in **C** so that $w(\alpha)$ must be divisible by n for all n, which implies that $w(\alpha) = 0$. That being so, one of the valuations w_∞ (the "point at infinity") is defined by $w_\infty(P) = -\deg P$ for all nonzero polynomials P. The others (the "points at finite distance") correspond one-to-one to the points $\zeta \in \mathbf{C}$: for all nonzero polynomials P, $w_\zeta(P)$ is the order of the zero ζ of P(X) (therefore, 0 if $P(\zeta) \neq 0$). The next step is to show that for each discrete valuation w of $\mathbf{C}(X)$, there are a finite number (at most equal to the degree $[K : \mathbf{C}(X)]$) of discrete valuations v_j on K, and, for each j, there is an integer $e_j \geqslant 1$ such that the restriction of v_j/e_j to $\mathbf{C}(X)$ is w. These "points" v_j are said to be *above* w. The points that are above w_∞ are said to be "points at infinity," the others, "points at finite distance." Indeed, the "abstract Riemann surface S," the set of "points" thus obtained, appears instantly as a "sheeted" structure recalling Riemann's conception.

5. The elements $f \in K$ such that $v(f) \geqslant 0$ for all the points v at finite distance are precisely those that are algebraic integers over the polynomial ring $\mathbf{C}[X]$. (Recall that this means that they are roots of an equation with coefficients in $\mathbf{C}[X]$, the leading coefficient being 1.) They form what is now called a *Dedekind ring* A, as do the algebraic integers over $\mathbf{Z}$ in an algebraic number field. During the years 1870–1880, Dedekind proceeded precisely to establish the theory of divisibility in algebraic number fields and now, in his memoir with Weber, he begins by similarly describing divisibility in A even before introducing the valuations. The maximal ideals $\mathfrak{P}_v$ of A correspond one-to-one to the points $v \in S$ at finite distance: $\mathfrak{P}_v$ is the set of $f \in A$ such that $v(f) > 0$. *Fractional ideals* of K are the A-modules $\mathfrak{a}$ contained in K for which there is a $c \neq 0$ in A such that $c\mathfrak{a} \subset A$; each of these can be written in a unique way as a product $\mathfrak{P}_{v_1}^{\alpha_1}\mathfrak{P}_{v_2}^{\alpha_2}\dots, \mathfrak{P}_{v_r}^{\alpha_r}$, where the v_j are points at finite distance and the α_j are positive or negative integers. This result can also be expressed by saying that $\mathfrak{a}$ is the set of the $f \in K$ such that:

(1) $$v_j(f) \geqslant \alpha_j \quad \text{for} \quad 1 \leqslant j \leqslant r$$

and

(2) $$v(f) \geqslant 0$$

for the other points $v \in S$ at finite distance.

6. However, as the behavior of the $f \in K$ "at infinity" plays no role in these definitions, they are manifestly insufficient for a deep study of K (or of S). So Dedekind and Weber generalize the notion of fractional ideal by introducing the notion of *divisor* on K: a divisor on K is, very simply, a family $D = (\alpha_v)_{v \in S}$ of positive or negative integers α_v such that $\alpha_v = 0$ except for a finite number of points (note that here v ranges over the *whole* of S). The structure of an *additive group* (isomorphic to $\mathbf{Z}^{(S)}$) is defined in a natural way on the set $\mathscr{D}(K)$ of divisors of K by setting $(\alpha_v) + (\beta_v) = (\alpha_v + \beta_v)$ and an *order* structure (compatible with the group structure) is defined whereby the relation $(\alpha_v) \leqslant (\beta_v)$ signifies that $\alpha_v \leqslant \beta_v$ for all $v \in S$. A divisor $D = (\alpha_v)$ such that $\alpha_v \geqslant 0$ for all $v \in S$ is said to be *positive* or *effective*, and every divisor can be written $D' - D''$, where D' and D'' are positive, (In fact, Dedekind and Weber give a name (*polygon*) only to the effective divisors and use differences of positive divisors throughout. Moreover, they use multiplicative notation in place of additive notation for the group of divisors.) The *degree*, deg(D), of $D = (\alpha_v)$ is the (positive or negative) integer $\sum_{v \in S} \alpha_v$; the *support* of D is the (finite) set of $v \in S$ such that $\alpha_v \neq 0$. The generalization of fractional ideal, defined

by conditions (1) and (2), is the set L(D), associated to each divisor $D = (\alpha_v)$, consisting of the $f \in K$ such that:

(3) $$v(f) \geqslant -\alpha_v \qquad \text{for } all\ v \in S,$$

(the negative sign is introduced for reasons of convenience). In fact, this is a *refinement* of the notion of fractional ideal because the set defined by (1) and (2) can be described as the union of the increasing family of the $L(D_m)$ where $D_m = (\alpha_v)$ is such that the α_v are equal to $-\alpha_j$ for $v = v_j$, to 0 for the other points at finite distance, and to the integer m for all the points at infinity of S. It is easy to show that L(D) is a *vector space* over **C** of *finite dimension, denoted* l(D).

7. The spaces L(D) had in fact been introduced by Riemann in a case of particular importance (V, 15). One of the first problems Riemann posed consisted in determining if there exists, on a Riemann surface of genus p, a rational function that has at most one pole of first order at m given points of S and is finite at the other points. If D is the divisor (α_v) where $\alpha_v = 1$ at each of these m given points and $\alpha_v = 0$ elsewhere, then the desired functions are those of L(D); note that L(D) contains the constants. The transcendental formula (7) of (V, 13), which expresses rational functions by means of integrals of the first and second kind, shows immediately that

(4) $$l(D) \geqslant \deg(D) - p + 1$$

("Riemann's theorem"), since $m = \deg(D)$ and the constants c_j are bound by $2p$ homogeneous linear relations. In particular, the inequality $m \geqslant p + 1$ implies that $\deg(D) - p + 1 \geqslant 2$; in other words, that there are *nonconstant* rational functions having no singularities other than at the m given points where they have at most a pole of order 1. On the contrary, if $m \leqslant p$, there are in general only such functions for *particular* positions of the m given points, and Roch (a student of Riemann), using the theory of adjoints (see VI, 17)), made explicit the conditions that the points must satisfy. One of the principal purposes of Dedekind and Weber's memoir is to obtain an expression for l(D) that recovers, in particular, the results of Riemann and Roch.

8. First of all, the conditions (3) can be interpreted in another way by means of the notion of *principal divisor*. For all $f \neq 0$ in K, there are only a finite number of valuations $v \in S$ such that $v(f) \neq 0$; denote by $(f)_0$ (resp. $(f)_\infty$) the positive divisor $(v(f)^+)$ (resp. $(v(f)^-)$. The degrees of these divisors are the same and are equal to the degree of the field K over the field $\mathbf{C}(f)$ generated by f, if f is not a constant. Set $(f) = (f)_0 - (f)_\infty$ and call (f) the principal divisor defined by f; thus $\deg((f)) = 0$. (In Riemann's "transcendental" interpretation, $(f)_0$ is the "divisor of zeros" and $(f)_\infty$ the "divisor of poles" of the rational function f; the fact that $\deg((f)) = 0$ is a consequence of the theorem of residues: df/f is integrated over the boundary of the simply connected part S′ of S, taking into account that each arc of this boundary occurs twice, in opposite directions.) Note, in particular, that if $v(f) \geqslant 0$ for all $v \in S$, then $f \in \mathbf{C}$ (only the constants are

everywhere holomorphic on a Riemann surface), and if, moreover, $v(f) > 0$ for some $v \in S$, then $f \neq 0$.

It follows that the relations (3) for $f \neq 0$ are equivalent to the inequality:

(5) $$(f) + D \geqslant 0.$$

9. The principal divisors form a subgroup $\mathscr{P}(K)$ of $\mathscr{D}(K)$ (isomorphic to the group $K^*/\mathbf{C}^*$ because, by the preceding remark, two elements of K^* that have the same principal divisor differ only by a constant factor). Two divisors belonging to the same class in the quotient group $\mathscr{C}(K) = \mathscr{D}(K)/\mathscr{P}(K)$ are said to be (linearly) *equivalent*; in other words, D and D′ are equivalent if and only if there exists $f \neq 0$ in K such that $D' - D = (f)$. It is clear that $\deg(D') = \deg(D)$ and $l(D') = l(D)$ for two equivalent divisors D, D′. Finally, $(f) + D = (g) + D$ for two elements f, g of $L(D)$ only if f/g is a constant of $\mathbf{C}$; in other words, the set $|D|$ of positive divisors equivalent to D can be identified with the projective space $\mathbf{P}(L(D))$ of dimension $l(D) - 1$.

10. As the "abstract Riemann surface" S defined by Dedekind and Weber has no topological structure, evidently it is not possible to define the notion of abelian integral in their theory. But they show that the notion of *meromorphic differential* on K can be defined in a purely algebraic way, and that this suffices to give an algebraic definition of the *genus* of K and to establish the Riemann-Roch theorem.

In the first place, the analytic concepts, for a rational function, of "value" at a point and of "derivative" must be recovered algebraically. For a point $v \in S$, the elements $f \in K$ such that $v(f) \geqslant 0$ form a subring $\mathfrak{o}_v$ of K, called the ring of the discrete valuation v; in $\mathfrak{o}_v$, the subset $\mathfrak{m}_v$ of f such that $v(f) > 0$ is the unique maximal ideal. The composite homomorphism $\mathbf{C} \to \mathfrak{o}_v \xrightarrow{\pi_v} \mathfrak{o}_v/\mathfrak{m}_v$ is bijective and $\mathbf{C}$ is identified with the field $\mathfrak{o}_v/\mathfrak{m}_v$ by this isomorphism. The *value* of $f \in \mathfrak{o}_v$ at v is the image $f_v = \pi_v(f)$ in $\mathbf{C}$ (for $f \notin \mathfrak{o}_v, f_v$ is, by convention, the symbol ∞ and v is said to be a pole of f); therefore, $v(f - f_v) > 0$, by definition. In $\mathfrak{o}_v$, there is always a *uniformizer* z (defined up to a unit of $\mathfrak{o}_v$), characterized by the relation $v(z) = 1$, and for all $f \in \mathfrak{o}_v$, an expression $f - f_v = uz^k$ with $k = v(f - f_v)$ and u invertible in $\mathfrak{o}_v$.

For nonconstant $f \in K$, the $v \in S$ such that $v(f - f_v) > 1$ or $v(1/f) > 1$ are called the *ramification points* of f, and the positive divisor $V(f) = (\alpha_v)$, where $\alpha_v = v(f - f_v) - 1$ for the v such that $v(f - f_v) > 1$, $\alpha_v = v(1/f) - 1$ for the v such that $v(1/f) > 1$, and $\alpha_v = 0$ at the other points of S, is called the *ramification divisor* of f.

11. Let z be a nonconstant element of K (i.e. $z \notin \mathbf{C}$); then, for every $f \in K$ there is an irreducible polynomial $F(X, Y) \in \mathbf{C}[X, Y]$, determined up to constant factor, such that $F(f, z) = 0$. The *derivative* of f with respect to z is defined as the element of K satisfying:

(6) $$\frac{df}{dz} = -F'_Y(f, z)/F'_X(f, z)$$

where F'_X and F'_Y are the usual partial derivatives. Dedekind and Weber show that it is equivalent to say

(7) $$\left(\frac{df}{dz}\right)_v = \left(\frac{f - f_v}{z - z_v}\right)_v$$

for infinitely many points $v \in S$. This allows them easily to see that the derivative thus defined satisfies the usual properties, particularly:

(8) $$\frac{dz_3}{dz_1} = \frac{dz_3}{dz_2} \cdot \frac{dz_2}{dz_1}$$

for three nonconstant elements z_1, z_2, z_3 of K. Moreover, they readily obtain the expression for the principal divisor:

(9) $$\left(\frac{dz_1}{dz}\right) = V(z_1) - 2(z_1)_\infty - (V(z) - 2(z)_\infty).$$

Setting $r(z) = \deg(V(z))$ and $n(z) = \deg((z)_\infty)$, the relation:

(10) $$r(z_1) - 2n(z_1) = r(z) - 2n(z)$$

for two arbitrary nonconstants $z, z_1 \in K$, immediately follows. Hence, the number

(11) $$g = \tfrac{1}{2}r(z) - n(z) + 1$$

is independent of the nonconstant z. By choosing a suitable basis of K over $\mathbf{C}(z)$, Dedekind and Weber show that g is an integer $\geqslant 0$ that is, by definition, the *genus* of K (or of S). (By transcendental methods, Riemann had obtained this formula expressing the genus of a Riemann surface in terms of the number of its branch points and the number of its sheets.)

12. Now, to define the differentials on K, consider the couples (f, z) where $f \in K$, $z \in K$, and $z \notin \mathbf{C}$. Define the product $u . (f, z)$ of one such couple by an element $u \in K$ as equal to (uf, z). In this set of couples, consider the relation $\sim$ where $(f, z) \sim (g, z_1)$ signifies that:

$$g = f\frac{dz}{dz_1}.$$

By (8) this is an *equivalence relation*, compatible with the multiplication by elements of K. The equivalence classes for this relation are, by definition, the *meromorphic differentials* on K. In particular, the class of $(1, z)$ is denoted dz, and, for each nonconstant z, every meromorphic differential can be written, uniquely, $\omega = f\,dz$ for some $f \in K$. If $\omega_1 = f_1\,dz$ and $\omega_2 = f_2\,dz$, the sum $\omega_1 + \omega_2$ is defined to be equal to $(f_1 + f_2)\,dz$. This addition is independent of the choice of z, since $f\,dz = (f(dz/dz_1))\,dz_1$ for all other nonconstants $z_1 \in K$ and, with the multiplication by elements of K, it makes the set Ω of differentials into a *vector space of dimension 1 over* K. For all $f \in K$, the differential $(df/dz)\,dz$ is denoted by df; it is also independent of the choice of the nonconstant z.

13. Now, a well-determined *divisor* $\Delta(\omega)$ can be associated to each differential $\omega \in \Omega$. If $\omega = dz$, where z is nonconstant, set $\Delta(dz) = V(z) - 2(z)_\infty$, and if $\omega = f\,dz$

with $f \in K$, set $\Delta(\omega) = (f) + \Delta(dz)$. From (9) it follows that if $\omega = f_1\,dz_1$ with z_1 nonconstant, then $(f) + \Delta(dz) = (f_1) + \Delta(dz_1)$, thus $\Delta(\omega)$ is intrinsically well defined; moreover, it follows from (11) that

(12) $$\deg(\Delta(\omega)) = 2g - 2.$$

All the divisors $\Delta(\omega)$ are equivalent by definition; they are called the *canonical* divisors of K and their class is the *canonical class*. For each $v \in S$, if z is a uniformizer for v, the component of $\Delta(f\,dz)$ at the point v (which is denoted $v(f\,dz)$) is simply $v(f)$ since $v(z) = 1$. Moreover, $\omega = f\,dz$ is said to be holomorphic at the point v if $v(\omega) \geqslant 0$; the holomorphic differentials ω (meaning: holomorphic at every point) for which $\Delta(\omega)$ is positive are said to be "of the first kind." The use of the same basis as that which served to define the genus allows Dedekind and Weber to prove that there exist g holomorphic differentials linearly independent over **C**; in other words, that

(13) $$l(\Delta(\omega)) = g.$$

14. Once these preliminaries are established, it is easy for Dedekind and Weber, always using the same special basis of K over $\mathbf{C}(z)$, to obtain the general form of the *Riemann-Roch theorem*:

(14) $$l(D) - l(\Delta - D) = \deg(D) + 1 - g$$

for an arbitrary divisor D, where Δ is any canonical divisor.

We will put aside the end of the Dedekind-Weber memoir, which is devoted to the algebraic definition of the notion of residue and of the notions of differentials of the second and of the third kind. Rather we mention here several consequences of the Riemann-Roch theorem due to various mathematicians about the same time.

It is convenient to designate by P a divisor (α_v) for which $\alpha_v = 0$ except for *one* valuation v_0 for which $\alpha_{v_0} = 1$; it is called, once again, a *point* and is identified with v_0. For every divisor D,

(15) $$l(D) \leqslant l(D + P) \leqslant l(D) + 1,$$

for if $D = (\beta_v)$ with $\beta_{v_0} = r$ and if z is a uniformizer at P, the linear map $f \mapsto (z^{r+1} f)_{v_0}$ of $L(D + P)$ to **C** is defined (since $v_0(f) \geqslant -r - 1$) and has kernel $L(D)$. By induction, it follows from (15) that if $D \leqslant D'$, then

(16) $$0 \leqslant l(D') - l(D) \leqslant \deg(D' - D).$$

If $\deg(D) < 0$, it is clear from the definition that $l(D) = 0$. If $\deg(D) \geqslant 0$, then

(17) $$l(D) \leqslant \deg(D) + 1$$

because if $n = \deg(D)$ and $D' = D - (n + 1)P$, then $\deg(D') < 0$ so that:

$$l(D') = 0 \qquad \text{and} \qquad l(D) \leqslant \deg(D - D') = n + 1$$

by (16).

These elementary remarks together with the theorem of Riemann-Roch (14) give the following results:

1° *If* $\deg(D) \geqslant 2g - 1$, *then*

(18) $$l(D) = \deg(D) + 1 - g$$

because in that case $\deg(\Delta - D) < 0$, therefore $l(\Delta - D) = 0$.

2° *If* $\deg(D) \geqslant 2g$, *then* $l(D - P) = l(D) - 1$ *for all points* P for it suffices to apply (18) to $D - P$.

3° *If* $l(D) > 0$ *and* $l(\Delta - D) > 0$, *then*

$$l(D) \leqslant \tfrac{1}{2}\deg(D) + 1$$

(Clifford's theorem). For, by replacing D and $D' = \Delta - D$ by equivalent divisors, it may be assumed that $D \geqslant 0$, $D' \geqslant 0$, and $\Delta \geqslant 0$. Moreover, by replacing D by $D - P$, if necessary, it may be assumed that $l(D - P) \neq l(D)$ for all points P. There is a $u \in L(D)$ such that $u \notin L(D - P)$ for all (the finite number of) $P \leqslant D'$. Then the map $L(D')/L(0) \to L(\Delta)/L(D)$, which to the class of f in $L(D')$ associates the class of fu, is injective; thus $l(D') - 1 \leqslant g - l(D)$, and it suffices to express $l(D') = l(\Delta - D)$ by the Riemann-Roch theorem.

15. Given a point P, denote the integer $l(r\,.\,P)$ by $N_r(P)$. It is clear (by 18)) that

$$1 = N_0(P) \leqslant N_1(P) \leqslant \cdots \leqslant N_{2g-1}(P) = g.$$

There are, therefore, exactly g integers $0 < n_1 < n_2 < \cdots < n_g < 2g$ for which there is *no* $f \in K$ whose divisor of poles is $n_j\,.\,P$ (or, in the analytic interpretation, no function on the Riemann surface having a single pole of order exactly equal to n_j at the point P). These integers are called the *Weierstrass integer gaps* at the point P. P is called a *Weierstrass point* if the sequence of the n_j is distinct from the sequence $(1, 2, \ldots, g)$; in other words, if

(19) $$\sum_{j=1}^{g} (n_j - j) > 0.$$

It can be shown that

(20) $$\sum_{P}\left(\sum_{j=1}^{g} (n_j - j)\right) = (g - 1)\,g(g + 1)$$

which means that there are only a *finite* number of Weierstrass points and that every field K of genus > 1 admits Weierstrass points. If $g > 1$ and if 2 is not an integer gap at P, the integer gap sequence is necessarily $(1, 3, 5, \ldots, 2g - 1)$; such a Weierstrass point is called *hyperelliptic*, and a field K for which all the Weierstrass points are hyperelliptic is said to be *hyperelliptic*. A field of genus 2 is always hyperelliptic.

2. The geometric school and linear series

16. Riemann's memoir on abelian functions avoided the use of geometric language, and it is only in 1863 that Roch and Clebsch begin to make the connection between Riemann's results and the projective geometry of plane algebraic

curves. Their initial success quickly attracted competitors, and, about 1870, an active school of birational geometry developed around Brill and Max Noether in Germany, Smith and Cayley in England, Halphen in France, Zeuthen in Denmark, and the first generation of Italian geometers, Cremona and Bertini. The principal theme of their research will be the mutual adaptation of the projective geometry from the beginning of the century and the new ideas of Riemann.

17. As the birational invariants of an algebraic curve are *a fortiori* projective invariants, a first problem was to try to express them as functions of the latter. In fact, by introducing "adjoints" (V, 15), Riemann had essentially given a formula expressing the genus of a plane curve as a function of its degree n and of the number r of its double points (assumed ordinary); however, his calculation is ill adapted to projective geometry because he starts with a curve defined by an equation $F(s, z) = 0$ in nonhomogeneous coordinates and introduces the degrees (not necessarily equal) of F with respect to s and z. M. Noether, assuming merely that the curve of degree n has only multiple points of orders r_j $(1 \leqslant j \leqslant s)$ with r_j distinct tangents ("*ordinary*" multiple points), obtains a more general formula for the genus:

$$g = \tfrac{1}{2}(n-1)(n-2) - \tfrac{1}{2}\sum_{j=1}^{s} r_j(r_j - 1). \tag{21}$$

18. Riemann seems not to worry about what happens when the singularities of the curve are more complicated; from his point of view, this is natural as only the "branches" intervene in the definition of the Riemann surface. The school of projective geometry could not adhere to this point of view, and it is quickly seen that the formula (21) gives, in general, a value *larger* than the genus when there is no hypothesis on the singular points; thus the right side of (21) is called the *virtual genus* (or *arithmetic genus*, cf. (VI, 46) and (VIII, 11)). In fact, the study of divisors and differentials on a curve having *arbitrary* singularities has been completed only very recently by Rosenlicht.

In the nineteenth century another path is followed; since it is a question of birational geometry, a plane curve can be replaced by any one of its transforms under a birational transformation, and if one of its transforms has only ordinary multiple points, it is possible to apply formula (21) and the theory of linear series (see (VI, 23)) to it. Therefore, the first problem on the agenda is the search for such birational transformations, or what is called the *resolution of singularities*.

19. It turns out that, in the course of his research in algebra, Kronecker had substantially obtained such transformations in 1858. This occurs in his memoir on the discriminant that he did not publish until 1880. Kronecker adopts here the starting point taken later by Dedekind-Weber: the field K of rational functions is considered as a finite extension of a subfield $\mathbf{C}(z)$ (isomorphic to $\mathbf{C}(X)$). If s generates K over $\mathbf{C}(z)$ and if:

$$F(s, z) = s^n + a_1(z)\, s^{n-1} + \cdots + a_n(z) = 0$$

is the irreducible equation satisfied by s, the points (s, z) satisfying both

$$F(s, z) = 0 \qquad \text{and} \qquad \frac{\partial F}{\partial s}(s, z) = 0$$

correspond, on the one hand, to the multiple points of the curve and, on the other, to the ramification points of the Riemann surface of s. Therefore, the corresponding values of z are obtained by solving the equation $\Delta_F(z) = 0$, where Δ_F is the discriminant of F. When s is replaced by another generator s_1 of K (which is necessarily of the form $s_1 = P(s, z)$, where P is a polynomial in s with coefficients in $\mathbf{C}(z)$), another irreducible equation:

$$F_1(s_1, z) = 0$$

is obtained. The discriminant Δ of the field K (over $\mathbf{C}(z)$) is the greatest common divisor (g.c.d.) of the discriminants of all these polynomials F_1, and the roots of $\Delta(z) = 0$ are the ramification points of the Riemann surface of any one of the functions s_1. There is an expression $\Delta_{F_1}(z) = \Delta(z)\,(G_1(z))^2$ where G_1 is a polynomial dependent on F_1, and Kronecker shows that the generator s_1 can be chosen so that G_1 has no root in common with Δ and has, moreover, only simple roots, which correspond to the double points with distinct tangents on the curve $F_1(s_1, z) = 0$. Therefore, the birational transformation $(s, z) \mapsto (s_1, z)$ accomplishes the "resolution of singularities" at least as far as the points at finite distance are concerned.

20. In 1871, Max Noether independently obtains the theorem of resolution of singularities for plane curves by another method that will remain fundamental to all the generalizations of this theorem to varieties of larger dimension ((VII, 44), (IX, 46)). The idea is to apply to the given plane curve $C \subset \mathbf{P}_2(\mathbf{C})$ a well-chosen succession of *quadratic transformations* (IV, 11). If the curve C of degree n has a multiple point O of order r that is not "ordinary," first of all, a quadratic transformation with base points O, A, B, where A and B are not on C, is made. The line AB meets C in n distinct points and each of the lines OA, OB meet C in $n - r$ points distinct from O. Then the transformed curve C' has degree equal to $2n - r$, and ordinary multiple points of order n, $n - r$, and $n - r$ at the points O, A, B, respectively. Moreover, AB meets C' in at most r points O'_i distinct from A and B. These points are again, in general, "nonordinary" multiple points of C'. Then it is seen that the virtual genus (21) of C' is strictly less than that of C unless the O'_i are simple points of C'. Therefore, if this procedure is iterated, it is seen that after a *finite* number of operations, the singularity of C at the point O will have been completely "resolved," meaning that the process has resulted in a curve having *fewer* "nonordinary" multiple points than C. Thus, by induction, a curve having only ordinary multiple points is finally obtained. Clebsch showed even that by one birational transformation (which, in this case, does *not* necessarily come from a birational transformation of $\mathbf{P}_2(\mathbf{C})$) a curve $C \subset \mathbf{P}_2(\mathbf{C})$, having only ordinary

multiple points, can be transformed into a plane curve having only double points with distinct tangents.

21. The preceding method led geometers of the nineteenth century to a closer analysis of singular points of plane curves: the appearance, after a quadratic transformation, of the points O'_i having multiplicities s_i on the curve C′ is expressed by saying that C has, in a "neighborhood of the first order" of O, multiple points O'_i of orders s_i that are "infinitely near" to O. The second operation gives, likewise, for each O'_i, points O_{ij} of multiplicities s_{ij} "infinitely near" O in a "neighborhood of the second order," and so on. Of course, it is necessary to show that these notions are independent of the choice of the quadratic transformations that define them; in fact, they can be directly defined from the Puiseux expansions of the diverse "branches" at O.

These notions are interesting because they permit the formulas established for ordinary multiple points to be extended naturally to the general case. For example, in M. Noether's formula (21), if the sum $\frac{1}{2}\sum_j r_j(r_j - 1)$ over the multiple points is replaced by the analogous sum over the multiple points and all their "infinitely near" multiple points of various orders, the value of the genus is obtained in every case.

22. Let C be an irreducible algebraic plane curve identified with its Riemann surface, and assume, at first, that C is nonsingular.* A *positive* divisor D on C (VI, 6) can be considered as a finite system of points on C each of which has, as coefficient, an integer $\geqslant 1$, its *multiplicity*. Given a divisor D_0 (positive or not) such that $l(D_0) = r + 1 > 0$, the functions $f \in L(D_0)$ can be written as restrictions to C of functions of the form:

(22)
$$\left(\sum_{j=0}^{r} \lambda_j P_j(x, y)\right) \Big/ Q(x, y)$$

in nonhomogeneous plane coordinates, where the P_j and Q are polynomials and the λ_j are arbitrary complex parameters. Any polynomial $F(x, y)$, not identically zero on C, restricts to a rational function (also denoted F by abuse of language) on C that is not identically zero. The divisor of zeros $(F)_0$ of this function is, since C is nonsingular, the set of points of intersection of C with the curve $F(x, y) = 0$, each point having its multiplicity as coefficient (in the sense of (IV, 8)). (Of course, the points at infinity common to the two curves must be included.) Therefore, the divisors $(f) + D_0$ corresponding to the $f \in L(D_0)$ can be written in the form $(\sum_j \lambda_j P_j)_0 - (Q_0) + D_0$. In other words, to each divisor D_0 such that $l(D_0) > 0$ is associated a variable family of positive divisors, defined as systems of points of intersection of C with a *linear* family of algebraic curves

(23)
$$\sum_{j=0}^{r} \lambda_j P_j(x, y) = 0.$$

* Translator's note: "sans singularités" is translated "nonsingular."

Conversely, suppose the polynomials P_j, linearly independent over $\mathbf{C}$ and not identically zero on C, are given. Consider the positive divisors $(\sum_j \lambda_j P_j)_0$ where the λ_j are variable complex numbers, not all zero. If D_0 designates one such, corresponding to some specified values λ_j^0 of the parameters, then $(\sum_j \lambda_j P_j)_0 = (f) + D_0$, where f is the rational function $(\sum_j \lambda_j P_j)/(\sum_j \lambda_j^0 P_j)$ on C. In other words, the divisors $(\sum_j \lambda_j P_j)_0$ form a *linear projective variety* in the projective space $|D_0|$. Such a variety of dimension r is called a *linear series of dimension r* on the curve; it is said to be *complete* when it is equal to the whole projective space $|D_0|$. Thus the study of linear series is essentially equivalent to the study of divisors.

23. It is, in fact, by *directly* defining linear series on a curve by the relation (23) that Brill and M. Noether give proofs of the algebraic results of Riemann (which Dedekind-Weber will prove later using the concept of divisor). From their point of view, this manner of approaching the theory has the advantage of staying in the framework of projective geometry (naturally, they must show the birational invariance of the notion of linear series); moreover, they can generalize the definition of linear series to plane curves with singular points. Most often, they restrict to the case where the curve C has only ordinary multiple points and they study, most particularly, the linear series obtained by intersecting C with a linear family (23) of *adjoint* curves of a fixed degree, that is, curves with the property that at a (ordinary) multiple point of C of order k, they have a multiple point of order at least $k - 1$. As we have seen, these curves were already introduced in Riemann's memoir (V, 15). For curves with ordinary multiple points and without singularities at infinity, Brill and Noether show that the integrals of the first or second kind have the form indicated by Riemann, the adjoint having to be of degree $\leqslant n - 3$ for an integral of the first kind, if the curve is of degree n.

For general values of the parameters λ_j in (23), the points of intersection of the curve (23) with C, which vary with the λ_j, are the points where the intersection number of the two curves is 1, and their number m is constant. They call the set of systems of such points on C a *linear series* of degree m and dimension r (often denoted g_m^r). One linear series is contained in another when each system of points of the first is contained in a system of points of the second. A linear series of fixed degree is said to be *complete* if it is maximal for this inclusion relation. Brill and Noether show that every complete linear series can be obtained by taking, for the $P_j(x, y) = 0$ in (23), adjoint curves of a fixed degree, possibly constrained to pass through a finite number of simple points of C; on the other hand, if C has genus $\geqslant 1$, they call the series defined by taking for the curves (23) *all* the adjoints of degree $n - 3$, the *canonical series*.

Finally, given two linear series, one defined by polynomials $P_j (0 \leqslant j \leqslant r)$, the other by polynomials $Q_h (0 \leqslant h \leqslant s)$, the *sum* of the two linear series is the linear series defined by the product polynomials $P_j Q_h$. With these definitions, Brill and Noether obtain results equivalent to those of Dedekind-Weber.

24. The same school (especially M. Noether and Halphen) also approaches the extension of the theory of plane algebraic curves to algebraic curves embedded in

an arbitrary projective space $\mathbf{P}_N(\mathbf{C})$. From the birational point of view, this is nothing new. Indeed, given an irreducible curve C in $\mathbf{P}_N(\mathbf{C})$ and a linear variety L of dimension $N - 3$ in $\mathbf{P}_N(\mathbf{C})$, which does not contain C, for every point P of C, aside from the finite number of points P where L meets C, there is a linear variety $L \vee P$ of dimension $N - 2$ passing through P and L. If L′ is a plane of $\mathbf{P}_N(\mathbf{C})$ not meeting L, $L \vee P$ meets L′ in a single point P′. In this way, a rational map (see (VI, 42)) $P \mapsto P'$ is defined from C onto the plane curve $C' \subset L'$; this map is called the *projection* of C onto C′ with *center* L. If L is in general position, it can be shown that this projection is a birational transformation, in other words, C and C′ are birationally equivalent. The problem is to obtain *projective* invariants (relative to the group of projective transformations of $\mathbf{P}_N(\mathbf{C})$) analogous to those that are introduced in Plücker's formulas (IV, 7), or to the multiplicities of the "infinitely near points" of a singularity of the curve (VI, 21). For example, if C is an ordinary nonsingular space curve ($N = 3$), projective invariants for C are immediately obtained from those of its projection C′ onto a plane from an arbitrary point in the space: degree, class of C′ (the number of tangents to C meeting an arbitrary line), the number of double points of C′ (called "apparent double points" of C, in other words, the number of chords of C passing through an arbitrary point), the number of inflection points of C′ (in other words, the number of osculating planes to C passing through an arbitrary point), and finally, the number of bitangents to C′ (in other words, the number of planes bitangent to C and containing an arbitrary point of the space). The genus of C (equal to that of C′) evidently can be deduced from them by (21); but a deeper study (Halphen) shows that not only are these numerical invariants insufficient to characterize the possible "types" of nonsingular space curves but it is even unlikely that any finite system of integer projective invariants will serve (in modern terms, the algebraic variety of curves for which these integers take the given values (see (VII, 31)) would not always be irreducible.)

The Brill-Noether theory can be applied also to curves embedded in a projective space $\mathbf{P}_N(\mathbf{C})$; it suffices to replace, in the definition of linear series, the plane curves (23) by a family of hypersurfaces:

$$(24) \qquad \sum_{j=0}^{r} \lambda_j F_j(x_0, x_1, \dots, x_N) = 0$$

(in homogeneous coordinates), where the F_j are homogeneous and of the same degree.

25. One of the most interesting consequences of the theory of linear series is that it gives a systematic means for defining, conversely, a birational transformation of an algebraic plane curve to an algebraic curve contained in a projective space $\mathbf{P}_N(\mathbf{C})$. Indeed, given a linear series on C defined by (23), to each point (x, y) of C, which is not one of the finite number of "fixed points" where all the P_j are zero, is associated the point of $\mathbf{P}_{r-1}(\mathbf{C})$ with homogeneous coordinates $P_j(x, y)$ $(1 \leqslant j \leqslant r)$; it is clear that this defines a rational map (see (VI, 42)) of C onto a curve

$C' \subset \mathbf{P}_{r-1}(\mathbf{C})$. It is possible that a point of C′ in general position corresponds only to a single point of C, in which case the linear series is said to be *simple*, and the transformation from C to C′ is birational; in the contrary case, the linear series is said to be *compound*: this means that, for every point P of C, all the systems of points of the given linear series that contain P contain also at least one other point that varies with P.

When the given linear series is simple and has no fixed points (except for the multiple points of C), the degree of C′ is equal to the number of points (other than the multiple points, and counted with their multiplicity) of a system of arbitrary points of the series (that is, to the *degree* of the corresponding divisors), the images in C′ of these systems of points being the intersections of C′ with the *hyperplanes* of $\mathbf{P}_{r-1}(\mathbf{C})$.

26. If C is a curve of genus g, then a complete linear series corresponding to a divisor D of degree $\geqslant 2g + 1$ is simple and has no fixed point (VI, 14). It can be shown that the curve $C' \subset \mathbf{P}_{r-1}(\mathbf{C})$ birationally equivalent to C, which is obtained by the preceding procedure, is *nonsingular*; if $g \geqslant 2$ and, for example, $D = 3\Delta$, where Δ is a canonical divisor, the resulting transformation is called the "tricanonical" embedding of C.

If $g \geqslant 3$ and if C is not hyperelliptic (VI, 15), these same assertions are again exact for $D = \Delta$. Thus a curve C′ of degree $2g - 2$ in the projective space $\mathbf{P}_{g-1}(\mathbf{C})$ is obtained; it is called the *canonical curve* corresponding to C (or the *canonical model* of the field of rational functions on C). Two nonhyperelliptic and birationally equivalent curves C_1, C_2 have canonical curves that are *projectively equivalent.*

27. Therefore, in principle, this result brings the problem of the quest for "birational invariants" of algebraic curves, introduced for the first time by Riemann under the name of "moduli," back to the classical theory of invariants (IV, 10). Riemann poses the problem of counting the number of (complex) parameters on which an isomorphism class of Riemann surfaces of genus $g \geqslant 2$ depends: if the number of sheets is n, the number of ramification points is $w = 2(n + g - 1)$ (VI, 11), then, once the points are fixed and n is fixed, the number of corresponding Riemann surfaces is finite. But two surfaces having n sheets and different ramification points can be isomorphic if the two systems of ramification points correspond to two functions generating the same field over $\mathbf{C}(z)$. However, if $n > 2g - 2$, any such function depends on $2n - g + 1$ parameters since its n poles can be taken arbitrarily, and still $n - g + 1$ parameters remain available, by Riemann's theorem (VI, 7). Finally, Riemann arrives at the number $2(n + g - 1) - (2n - g + 1) = 3g - 3$ parameters that he calls "moduli." Except in particular cases, no result more precise on what is meant by the "variety of moduli" has been obtained until very recently.

28. A curve C of $\mathbf{P}_N(\mathbf{C})$ can be projected, not now on a plane but on a linear variety of dimension 3, so that if C is nonsingular then so is its projection. In

conjunction with (VI, 26), this shows that every algebraic curve is birationally equivalent to a nonsingular curve *in* $\mathbf{P}_3(\mathbf{C})$. On the contrary, as a nonsingular plane curve is of genus $\frac{1}{2}(n-1)(n-2)$ by (21), for values of the integer g that are not of this form (for example, already for $g = 2$), it follows that there is no nonsingular plane curve of genus g.

In general, for a curve C of degree n and genus g in a projective space $\mathbf{P}_r(\mathbf{C})$, there are relations of inequality between the three numbers n, g, r. For example (Castelnuovo), if e is the integer satisfying the inequalities

$$\frac{n-1}{r-1} - 1 \leqslant e < \frac{n-1}{r-1},$$

then:

$$g \leqslant e(n - r - \tfrac{1}{2}(e-1)(r-1)).$$

Every curve of genus g is birationally equivalent to a curve of degree $g - [\frac{g}{3}] + 2$.

29. Hyperelliptic curves exhibit special behavior. For a fixed genus $g \geqslant 2$, the isomorphism classes of these curves depend only on $2g - 1$ "moduli." A hyperelliptic curve of genus g is birationally equivalent to a plane curve of degree $g + 2$ having an ordinary multiple point of order g, and also to a curve with (nonhomogeneous) equation $y^2 = P(x)$, where P is a polynomial of degree $2g + 2$ having only simple roots.

Curves of genus 0 are called *rational* or *unicursal*; they are birationally equivalent to the line $\mathbf{P}_1(\mathbf{C})$. For each integer n, the "normal" curve of degree n in $\mathbf{P}_n(\mathbf{C})$, defined by the parametric representation:

$$x_0 = 1,\ x_1 = t,\ x_2 = t^2,\ \ldots,\ x_n = t^n, \tag{25}$$

is, up to projective equivalence, the only curve of degree n in $\mathbf{P}_n(\mathbf{C})$ not contained in a hyperplane, and every rational curve of degree n in $\mathbf{P}_r(\mathbf{C})$ with $r < n$ is a projection of the "normal" curve. Therefore, there are, in this case, no "moduli" in Riemann's sense. Every curve of genus 1 is birationally equivalent to a nonsingular plane cubic. To such a cubic can be attached, as was seen in (IV, 6), the projective invariant equal to the cross ratio of the four tangents to the curve issuing from one of its points; it can be proved to be also a birational invariant, the unique "modulus" characterizing the isomorphism classes of curves of genus 1. The principal interest of nonsingular curves of genus 1 rests in the fact that they naturally have the structure of an algebraic group (see (VII, 57)).

3. The "transcendental" theory of algebraic varieties

30. From 1870, Cayley, Clebsch, and M. Noether try to develop the theory of algebraic surfaces (in $\mathbf{P}_3(\mathbf{C})$) by studying, as Riemann did for curves, the double integrals:

$$\iint R(x, y, z)\, dx\, dy \tag{26}$$

over an algebraic surface $F(x, y, z) = 0$ (the equation in nonhomogeneous coordinates), where R is a rational function, and z is assumed replaced by the "implicit function" of x and y deduced from the equation of the surface. From 1885, Picard intensified the study by joining to it the simple integrals:

$$\int P(x, y, z)\, dx + Q(x, y, z)\, dy,$$

where P and Q are rational functions such that the differential $P dx + Q dy$ is exact (in other words,

$$\frac{\partial P}{\partial y}(x, y, z) = \frac{\partial Q}{\partial x}(x, y, z)$$

on the surface), and, as for (26), z must be considered as an "implicit function" of x and y.

If, already, the clear conception of algebraic curves in $\mathbf{P}_2(\mathbf{C})$ and abelian integrals on such a curve present some difficulty for the "intuition," since (even supposing the curve to be nonsingular) it is a question of a variety of 2 real dimensions in a space of 4 real dimensions, how much more reason is there to give precise analytic definitions when it deals with algebraic surfaces or of algebraic varieties of higher dimension. The definition of a simple integral $\int P(x, y, z)\, dx$ closely imitates that of abelian integrals: for each piecewise differentiable map $t \mapsto (x(t), y(t), z(t))$ of an interval $[a, b]$ of $\mathbf{R}$ into the surface S under consideration, the integral over the image of this interval in S (an "oriented singular 1-simplex") is by definition the complex number:

$$\int_a^b P(x(t), y(t), z(t))\, x'(t)\, dt.$$

In the same way, to define (26) it is necessary to consider a piecewise differentiable map:

$$(u, v) \mapsto (x(u, v), y(u, v), z(u, v))$$

of a triangle $T \subset \mathbf{R}^2$ into S, and the integral over the image of T (an "oriented singular 2-simplex") is the complex number:

$$\iint_T R(x(u, v), y(u, v), z(u, v)) \frac{\partial(x, y)}{\partial(u, v)}\, du\, dv.$$

From this, the notion of a simple (resp. double) integral over a 1-chain (resp. 2-chain), that is, over a formal linear combination of oriented singular 1-simplices (resp. 2-simplices) with coefficients in $\mathbf{Z}$ (or in $\mathbf{R}$, or in $\mathbf{C}$), is deduced in a straightforward way, and it is easy to generalize these notions to any number of dimensions whatsoever.

31. Another important preliminary point to elucidate before it is possible to construct a theory of simple or double integrals on an algebraic surface is the influence of the singularities of the surface on the behavior of these integrals. A superficial examination already shows that the singular points of an algebraic

surface present problems of classification that are much more serious than those of algebraic curves: a surface can contain curves all points of which are singular for the surface ("singular curves"), as well as isolated singular points of very diverse nature. The notion of "branch" of a curve generalizes only in a much more complicated fashion: in general, there are no longer expansions, analogous to the Puiseux expansions, which, for each "branch" at a singular point of the surface, would give the coordinates in convergent series of two complex parameters u, v; such series only can be obtained that converge, not in a neighborhood of (0, 0) in $\mathbf{C}^2$, but in regions defined by inequalities of the type $|u| < r$, $|v| < r'|u|^k$. Thus, this seems to exclude the possibility of arguing, as Riemann did, on a nonsingular variety where the points would be "branches" of the given surface (see (VII, 44)).

32. Another route is to try to "resolve the singularities" as for algebraic curves, that is, to obtain a birational transformation of the given surface S to a nonsingular surface S′ contained in some suitable $\mathbf{P}_N(\mathbf{C})$. Since M. Noether, numerous attempts have been made to demonstrate the existence of such a "birational model" S′ in $\mathbf{P}_5(\mathbf{C})$. The remark that, by a quadratic transformation, an isolated singular point can "blow up" into a singular curve, for example, illustrates the difficulties to be surmounted. The first proof, secure from all criticism, was given only in 1935 by R. Walker (see VII, 44)).

However, it is necessary to note here the fundamental difference with the theory of algebraic curves: for an algebraic curve, a "nonsingular model" is essentially *unique*, that is, two such models can be transformed into one another by a *bijective* birational map (therefore everywhere defined as is its inverse, contrary to what happens for general birational transformations (cf. (IV, 11)). On the contrary, two nonsingular algebraic surfaces can very well be birationally equivalent without being able to be transformed one into the other in this way: an example is given by the projective plane $\mathbf{P}_2(\mathbf{C})$ and a nondegenerate quadric Q in $\mathbf{P}_3(\mathbf{C})$, which can be identified with $\mathbf{P}_1(\mathbf{C}) \times \mathbf{P}_1(\mathbf{C})$, and $\mathbf{P}_2(\mathbf{C})$ and $\mathbf{P}_1(\mathbf{C}) \times \mathbf{P}_1(\mathbf{C})$ are not even homeomorphic (the homology in dimension 2 is different).

This essential fact explains why it is not possible, on the model of the Dedekind-Weber theory, to take into consideration only the field of rational functions on an irreducible algebraic surface; in other words, many results that are invariant under *every* birational transformation (or, so-called, *absolute invariants*) must not be expected. The birational geometry of surfaces must be satisfied with "relatively invariant" notions, that is those that are preserved by a bijective, birational transformation (what will be called later (VIII, 25) an *isomorphism* of algebraic varieties).

33. After M. Noether and Picard, most of the work on algebraic surfaces, in the period we are considering, adopted a point of view analogous to that of Brill and Noether, namely, to restrict to surfaces embedded in $\mathbf{P}_3(\mathbf{C})$ having only "ordinary" singularities. By this is meant surfaces that can be obtained from a

nonsingular surface embedded in $\mathbf{P}_5(\mathbf{C})$ by *projection* (VI, 24) from a linear projective variety L in this space. If such a nonsingular surface is projected into a space $\mathbf{P}_4(\mathbf{C})$ from a variety L reduced to a single point in general position, a surface in $\mathbf{P}_4(\mathbf{C})$ having only isolated singular points is obtained. By projecting into a space $\mathbf{P}_3(\mathbf{C})$ from a line L in general position, a surface S is obtained whose singular points are the points of a *double curve* Γ having a finite number of triple points with distinct tangents that are also triple points of S. At the simple points of Γ, there are two distinct tangent planes to the two "nappes" of the surface at this point except for a finite number of "pinch-points" where the two tangent planes coincide (it can have there many types of pinch-points, one of which is qualified as "ordinary," and it may always be assumed that there are no other types). Moreover, there are relations, due to Salmon, Cayley, and Zeuthen, that are analogous to the Plücker formulas, between the projective invariants of Γ (VI, 24) and four projective invariants of S: its degree, its rank (the class of a general plane section of S), its class (the number of tangent planes through a general line), and the number of pinch-points on Γ.

For each point of an algebraic surface in $\mathbf{P}_3(\mathbf{C})$ having ordinary singularities, there is again a decomposition into "branches" with Puiseux expansions in two variables for the coordinates in each branch, coming from the "projection" from the nonsingular model, which consequently plays the same role in the study of the topology of the algebraic surface and the integrals that are attached to it as the Riemann surface for algebraic curves.

34. In the series of memoirs on simple and double integrals on algebraic surfaces that he published over a period of 20 years, Picard introduced a multitude of ideas and original methods that were to have considerable influence on later developments. The limits of this work do not permit us to analyze these works in detail. Attacking a completely unexplored subject, Picard uses techniques innately very diverse and not even well dissociated during this period: the theory of analytic functions, algebraic topology, then, a little later, the theory of linear systems due to the Italians (that was to become, in our time, the cohomology of sheaves). We will have the occasion to mention Picard's contributions to the development of each of these techniques when we examine them.

35. However, we mention, forthwith, a general method inaugurated by Picard that is essentially a recurrence on dimension. Given a surface with (nonhomogeneous) equation $F(x, y, z) = 0$, he considers the sections, by the planes $y = \text{constant}$, of this surface that are, in general, irreducible and of the same genus p. He applies Riemann's theory to these curves, studying the manner in which the corresponding abelian integrals depend on the parameter y. In particular, the $2p$ periods of the integrals of the first kind on the curve satisfy, as functions of y, a homogeneous linear differential equation of order $2p$ called the Picard-Fuchs equation. Picard uses it in an ingenious way, and, after having been abandoned for numerous years, this equation has regained much importance in recent work.

36. At the beginning of Picard's research, the concepts of algebraic topology remained almost at the point where Riemann had left them, founded on "intuitive" arguments concerning the position and the deformations of curves or surfaces, which, to say the least, are difficult to follow in spaces of dimension $\geqslant 4$. Thrust by his work up against problems of *analysis situs* (as it was called then) H. Poincaré decided, in 1900, to give this branch of mathematics a solid foundation by inventing the technique of "simplicial complexes" that is applicable to triangulable varieties. Recall that, in its present form, this technique consists in associating to each n-chain on a triangulated variety M, its boundary, which is an $(n-1)$-chain. This defines a homomorphism ∂ from the group of n-chains to the group of $(n-1)$-chains (with coefficients in $\mathbf{Z}$ or $\mathbf{Q}$ or $\mathbf{R}$ or $\mathbf{C}$), and $\partial \circ \partial = 0$. The *nth homology group* $H_n(M, \mathbf{Z})$ (or $H_n(M, \mathbf{Q})$ or $H_n(M, \mathbf{R})$ or $H_n(M, \mathbf{C})$) is the quotient of the kernel of ∂ in the group of n-chains (the elements of the kernel are called *n-cycles*) by the image of ∂ (*boundaries* of the $(n+1)$-chains). The nth Betti number R_n or M is the dimension of the vector space $H_n(M, \mathbf{R})$. It is customary to consider the direct sum of the groups of n-cycles (for every integer $n \geqslant 0$) and to call a *cycle* a sum of n-cycles for distinct values of the integer n; the dimension of the cycle is then the largest value of n for which the component of order n of the cycle is $\neq 0$. The direct sum $H_\bullet(M, \mathbf{Z})$ (resp $H_\bullet(M, \mathbf{Q})$, $H_\bullet(M, \mathbf{R})$, $H_\bullet(M, \mathbf{C})$) of the $H_n(M, \mathbf{Z})$ (resp. $H_n(M, \mathbf{Q})$, $H_n(M, \mathbf{R})$, $H_n(M, \mathbf{C})$) is defined in the same way. Then the n-cycles and their homology classes are called *homogeneous* (of dimension n). A 0-cycle with integer coefficients is simply a formal linear combination with (positive or negative) integer coefficients of points of M; the sum of these coefficients is called the *degree* of the 0-cycle; it is the same for two homologous 0-cycles.

Another important notion from Poincaré's theory (that Lefschetz later made precise and developed) is that of "intersection product" $C'.C''$ of a p-chain C' and a q-chain C'' in M where $p+q \geqslant n = \dim(M)$. If C' and C'' are in general position with respect to one another, a $(p+q-n)$-chain $C'.C''$ can be defined so that it is a *cycle* if C' and C'' are cycles; its homology class depends only on the classes of C' and C'', and

$$C''.C' = (-1)^{(n-p)(n-q)}\, C'.C''.$$

If $p+q=n$, the case where the dimensions of C' and C'' are said to be *complementary*, the degree of the 0-cycle $C'.C''$ is denoted $(C'.C'')$ and is called the *intersection number* of C' and C''.

Thus the structure of a noncommutative ring is defined on $H_\bullet(M, \mathbf{Z})$ (resp. $H_\bullet(M, \mathbf{Q})$, $H_\bullet(M, \mathbf{R})$, $H_\bullet(M, \mathbf{C})$); this ring is called the *homology ring* of M on $\mathbf{Z}$ (resp. $\mathbf{Q}$, $\mathbf{R}$, or $\mathbf{C}$). It is customary to calculate directly on the cycles instead of their homology classes, by writing $A \underset{h}{\sim} B$ to signify that the classes of the cycles A and B in the homology ring are the same: then A and B are said to be *homologically equivalent* (or simply, *homologous*). Note that the product A . B of *any* two cycles can then be defined (up to homological equivalence) as a cycle whose class in the homology ring is the product of the classes of A and B; this can be interpreted geometrically by saying that it is a cycle equivalent to a product $A'.B'$, where A'

and B′ are "in general position" and are obtained from A and B by an arbitrarily small "deformation."

37. This homological technique is applicable to algebraic varieties (with singularities) in a $\mathbf{P}_N(\mathbf{C})$ because these varieties are *triangulable*. Poincaré had already sketched a proof of this fact, which was completely proved by van der Waerden in 1930. Moreover, the homology of algebraic varieties exhibits remarkable features. First of all, an algebraic subvariety of $\mathbf{P}_N(\mathbf{C})$ is always *orientable*, and an irreducible variety V of (complex) dimension n is a $2n$-*cycle* (and not only a $2n$-chain), because the singularities of V are in a subcomplex of V of real dimension $2n - 2$ (and not $2n - 1$). Furthermore, around 1920, Lefschetz made the elementary (but essential) observation that if V is a nonsingular algebraic variety of complex dimension n in $\mathbf{P}_N(\mathbf{C})$ and if C′ and C″ are two irreducible algebraic subvarieties of V of complementary dimensions p, q $(p + q = n)$ and in "general position" (which means, in this context, that each of the intersection points of C′ and C″ is nonsingular on C′ and C″ and that the intersection of the tangent planes at this point on C′ and C″ is reduced to this point), then the coefficient of each of the points of the intersection $C' \cap C''$ in the calculation of the 0-cycle $C'.C''$ is the same, equal to $+1$ for appropriate orientations of C′ and C″. Therefore, the intersection number $(C'.C'')$ in the sense of algebraic topoplogy is, in fact, the *number of points* of $C' \cap C''$.

We observe that the number $(C'.C'')$ is defined even if C′ and C″ are not in "general position" but, in this case, it can be *negative*, which simply means that there are no *algebraic* cycles C_0', C_0'' in general position homologous respectively to C′ and C″. For example, if D is a line on a nonsingular cubic surface V and C is the conic where the plane passing through D cuts V, $D + C$ is homologous to every intersection L of V by a plane (it is deduced from it by continuous variation); therefore, $((D + C).D) = (L.D) = 1$ and as $(C.D) = 2$, it follows that $(D.D) = -1$.

38. As we have noted above, part of Picard's studies on simple or double integrals on an algebraic surface is, in fact, a study of the homology of the surface. Around 1920, Lefschetz revived and developed Picard's arguments using Poincaré's simplicial technique and obtained very precise information on the homology of a nonsingular irreducible algebraic surface $S \subset \mathbf{P}_3(\mathbf{C})$ with equation $F(x, y, z) = 0$. If C_y is the curve that is the inverse image of y in S under the projection pr_2 on the second coordinate axis, then C_y is a nonsingular irreducible curve of constant genus p except for a finite number of points y_j $(1 \leqslant j \leqslant \mu)$; the C_{y_j} acquire a unique double point with distinct tangents that lowers the genus by one. Picard and Lefschetz consider an arbitrary 1-cycle γ on a curve C_a (a distinct from the y_j) and make it vary continuously with y in C_y; they show that when y travels around a y_j in $\mathbf{C}$ describing a loop with origin a, the cycle γ becomes $\tau_j(\gamma)$, homologous to $\gamma + (\gamma.\delta_j)\delta_j$, where δ_j is a 1-cycle of C_a whose class in $H_1(C_a, \mathbf{Z})$ is well determined. Moreover, δ_j is an "evanescent" cycle in the sense that if a is joined to y_j by an arc

l_j, there is a 2-chain Δ_j in $\mathrm{pr}_2^{-1}(l_j)$ such that $\partial\Delta_j = \Delta_j \cap C_a = \delta_j$; $\Delta_j \cap C_y$ is a 1-cycle deduced from δ_j by continuous variation for $y \in l_j$, and $\Delta_j \cap C_{y_j}$ is the double point of C_{y_j}. Then Lefschetz shows how the homology groups $H_i(S, \mathbf{Z})$ for $i = 1, 2, 3$ can be described using the chains δ_j, Δ_j and a base $(\gamma_k)_{1 \leqslant k \leqslant 2p}$ of the homology of the curve C_a. In particular, there are $2p - R_1$ homologically independent cycles δ_j on C_a.

Using similar methods, Lefschetz took up the general study of the homology of an irreducible variety V (contained in a $\mathbf{P}_N(\mathbf{C})$) of any (complex) dimension n. He obtained a whole series of remarkable results to which we will return in Hodge theory (see (VII, 10)). We mention, in particular, the general expression of Picard's method of recurrence, namely, that for a section H of V by a hyperplane in general position, the canonical homomorphism $H_i(H, \mathbf{Z}) \to H_i(V, \mathbf{Z})$ induced by the injection $H \to V$ is bijective for $0 \leqslant i \leqslant n - 2$ and surjective for $i = n - 1$. Another general result of Lefschetz concerns particular properties of the Betti numbers of algebraic varieties: for Betti numbers of even dimension, $R_{2p} > 0$; for $p \leqslant n$, $R_p \geqslant R_{p-2}$; and, finally, R_{2p+1} is, necessarily, an even number (but can be 0).

We will also have the occasion to speak of the results of Picard, Poincaré, and Lefschetz with regard to the characterization of algebraic 2-cycles (VI, 37) and their relation with double integrals ((VI, 48) and (VII, 53)), and when we examine their generalization in the cohomology of sheaves (IX, 85 to 102).

39. Finally, we indicate how the homological study of algebraic surfaces has clarified and simplified the theory of *correspondences* between algebraic curves. We have seen (IV, 12) how the school of projective geometry had introduced and used the notion of (m, n) correspondence on a line (or on a rational curve). It was not difficult to generalize this definition on a curve of any genus g but, as early as 1866, Cayley noticed that Chasles's formula giving the number of united points of the correspondence as equal to $m + n$ was no longer correct, in general, and he concluded, semi-empirically, that $m + n$ must be replaced by $m + n + 2gv$, where v is an integer (positive, negative, or zero) depending on the correspondence, and called its "valence." A little later, Brill gave proofs of this formula in a certain number of cases. The "valence" v, when it exists, is easily defined using the notion of divisor: it is the integer (necessarily unique if $g \geqslant 1$) such that, when $P'_1, \ldots, P'_m$ designate the m points corresponding to a point P on the curve, all the divisors $v \cdot P + P'_1 + \cdots + P'_m$ (when P varies) are equivalent. For example, if to a point P on a plane curve C of degree n, correspond the $n - 2$ points $P'_j (1 \leqslant j \leqslant n - 2)$, other than P where the tangent at P meets C, then v can be taken to be 2; the divisors $2P + \sum_{j=1}^{n-2} P'_j$ belong to the linear series of the intersections of C and of the lines of the plane.

40. In 1886, Hurwitz took up the question, under a more general form, by using the theory of abelian integrals, and Lefschetz showed in 1927 that his results can be presented in purely topological form, where the notion of integral no longer

intervenes. In general, consider two nonsingular irreducible algebraic curves C_1, C_2 of genus g_1, g_2. A *correspondence* (or *effective* correspondence) between C_1 and C_2 is, by definition, an algebraic curve Γ (irreducible or not) in the surface $C_1 \times C_2$; it is *nondegenerate* if it contains no irreducible component of the form $P_1 \times C_2$ or $C_1 \times P_2$ (P_1 a point of C_1, P_2 a point of C_2). Then the intersection numbers $m_1 = (\Gamma . (P_1 \times C_2))$ and $m_2 = (\Gamma . (C_1 \times P_2))$ are well defined and independent of $P_1 \in C_1$ and $P_2 \in C_2$. It is easy to obtain a base for the homology of $C_1 \times C_2$ by taking the products of the cycles of a base for the homology of C_1 and the cycles of a base for the homology of C_2. As Γ is a 2-cycle on $C_1 \times C_2$, it is, therefore, homologous to a unique linear combination, with integer coefficients, of this base.

41. The essential fact is that the correspondence Γ defines two *homomorphisms* of homology groups:

$$\Gamma_* : H_1(C_1, \mathbf{Z}) \to H_1(C_2, \mathbf{Z})$$

$$\Gamma'_* : H_1(C_2, \mathbf{Z}) \to H_1(C_1, \mathbf{Z})$$

where, for example, Γ_* maps the class of a 1-cycle γ_1 of C_1 to the class of the projection on C_2 of the intersection 1-cycle $\Gamma . (\gamma_1 \times C_2)$ on $C_1 \times C_2$; Γ'_* is defined in the same way by exchanging the roles of C_1 and C_2.

That being so, if $\Gamma^{(1)}$ and $\Gamma^{(2)}$ are two nondegenerate correspondences between C_1 and C_2, the *number of coincidences* of these two correspondences is, by definition, the intersection number $(\Gamma^{(1)} . \Gamma^{(2)})$ of the 2-cycles $\Gamma^{(1)}$ and $\Gamma^{(2)}$ on $C_1 \times C_2$; Lefschetz's general theory for the calculation of this number in the theory of simplicial complexes gives the formula of Hurwitz:

(27) $$(\Gamma^{(1)} . \Gamma^{(2)}) = m_1^{(2)} m_2^{(1)} + m_1^{(1)} m_2^{(2)} - \operatorname{Tr}(\Gamma_*^{(2)} \circ \Gamma_*^{(1)\prime}).$$

When $C_1 = C_2 = C$, the calculation of the number of united points of a correspondence Γ reduces to that of the number of coincidences of Γ and of the "identity correspondence" Δ (the diagonal of $C \times C$), and the formula:

(28) $$(\Gamma . \Delta) = m_1 + m_2 - \operatorname{Tr}(\Gamma_*),$$

due also to Hurwitz, is obtained.

The case of correspondences with "valence" corresponds to the case where the endomorphism Γ_* of $H_1(C, \mathbf{Z})$ is an integer multiple $-v . 1_{H_1(C,\mathbf{Z})}$ of the identity, in which case (28) gives the Cayley-Brill formula again.

There are examples of correspondences where there is no "valence," but for "general" values of the "moduli" of a curve C, it can be proved that every correspondence on C is a correspondence with valence.

42. We observe here that there does not exist necessarily, on a given curve C, a (m, n) correspondence on $C \times C$ for given m and n. In general, with the notation of (VI, 40), let d_1 (resp. d_2) be the number of points of C_1 (resp. C_2) such that at least two of the points of C_2 (resp. C_1) corresponding to it coincide. Then, there is Zeuthen's formula:

(29) $$m_1(2g_1 - 2) + d_1 = m_2(2g_2 - 2) + d_2.$$

In particular, if $m_1 = 1$ (in which case the correspondence is said to be *rational*), then, necessarily, $d_1 = 0$. If, moreover, $g_2 = g_1$, then:

$$2g_1 - 2 = m_2(2g_1 - 2) + d_2,$$

and, as $d_2 \geqslant 0$, this is possible (for $g_1 > 1$) only if $m_2 = 1$. In other words, a *rational* correspondence between two curves of the same genus > 1 is necessarily *birational* (Weber's theorem). If $m_1 = 1$ and $g_1 = 0$, then $-2 = m_2(2g_2 - 2) + d_2$, which is possible only if $g_2 = 0$. In other words, if there is a rational correspondence between a rational curve C_1 and a curve C_2, then C_2 is also rational (Lüroth's theorem); similarly, if $g_1 = 1$ then necessarily $g_2 = 0$ or $g_2 = 1$.

Finally, it can be shown that if C is a curve of genus $g \geqslant 2$, there exists only a *finite* number of birational correspondences on C (the theorem of Schwarz-Klein); Hurwitz even proved that this number is $\leqslant 84(g - 1)$.

4. The Italian School and the Theory of Linear Systems

43. Let $V \subset \mathbf{P}_N(\mathbf{C})$ be a nonsingular irreducible algebraic variety, K its field of rational fractions (which can be defined as the field of restrictions to V of rational functions on $\mathbf{P}_N(\mathbf{C})$, defined at, at least one point of V). The definition of divisors given in (VI, 6) generalizes without difficulty, except that here the model V and not the field K is the starting point; if V is of dimension n, the set S of the subvarieties W of V that are irreducible and of dimension $n - 1$ (or, as is also said, of *codimension 1*) is considered. A *divisor* $D = (\alpha_W)_{W \in S}$ is a family of positive or negative integers, zero except for a finite number of W. On the set $\mathscr{D}(V)$ of divisors, the structure of an ordered group is defined, as well as the notions of degree and of support (the support being the union of the subvarieties W of codimension 1 such that $\alpha_W \neq 0$). For each subvariety $W \in S$, there is a well-determined discrete valuation v_W of K; the functions $f \in K$ such that $v_W(f) \geqslant 0$ being those defined at all the points of W, and the $f \in K$ such that $v_W(f) > 0$ being those that are zero on W (but the v_W do not constitute *all* the discrete valuations on K). For each nonconstant function $f \in K$, the principal divisor (f) of f is defined as in (VI, 8), as is the vector space L(D), for each divisor D, consisting of the $f \in K$ such that $(f) + D \geqslant 0$; as before, it is of finite dimension $l(D)$, and the set $|D|$ of positive divisors linearly equivalent to D is a projective space of dimension $l(D) - 1$.

44. If V is an n-dimensional nonsingular algebraic variety, a *meromorphic differential n-form* ω on V can be expressed, in the neighborhood of each point of V, in the form $f(u_1, \ldots, u_n)du_1 \wedge du_2 \wedge \cdots \wedge du_n$, where the $u_j \in K$ are the local (complex) coordinates at the point under consideration, and $f \in K$. For every other system of local coordinates (u_j') at the same point:

$$\omega = f' du_1' \wedge du_2' \wedge \cdots \wedge du_n',$$

where f'/f is defined and nonzero in the neighborhood of the point under consideration. For each irreducible subvariety $W \in S$, and for a system of local coordinates defined at a point of W, the integer $v_W(f)$ is independent of this system of local coordinates; it is denoted $v_W(\omega)$ and is shown to be zero except for a finite number of $W \in S$. The divisor $\Delta(\omega)$ is defined to be equal to $(v_W(\omega))_{W \in S}$. Since two meromorphic differential n-forms differ only by a factor in **K**, all the divisors $\Delta(\omega)$ are linearly equivalent, and their class is called, as before, the *canonical class* of divisors on V.

45. Finally, as in (VI, 22), the notion of divisor is connected to that of *linear system of subvarieties of codimension 1*, defined as a projective space made up of the divisors of zeros $(\sum_j \lambda_j P_j)_0$ of a linear family of linearly independent polynomials $\sum_j \lambda_j P_j$. It is from this aspect that, at first for the case of algebraic surfaces (most often embedded in $\mathbf{P}_3(\mathbf{C})$), the same geometers who, around 1870, developed the study of algebraic curves using linear series, try to find the analogues of the results of this study for algebraic varieties of dimension $\geqslant 2$ without, moreover, dissociating their methods from the "transcendental" methods of which we spoke above. Once again, as for Picard's work, it will be impossible for us to follow in detail the often extremely complex notions introduced and discussed in the course of this period; we will be content to indicate the principal directions of the research of the geometers of this school, returning later to their results when they could be translated, in the modern age, into a more accessible language.

46. The first numerical invariant attached to a nonsingular algebraic surface (or to a surface with ordinary singularities) is the *geometric genus* p_g, defined by Clebsch, by analogy with the genus of a curve, as equal to the maximum number of linearly independent double integrals (26) that are "of the first kind;" that is, those that always have finite value over an arbitrary singular 2-simplex. If the surface S under consideration is nonsingular, it is easy to see that this condition is equivalent to saying that the meromorphic 2-form $\omega = R(x, y, z)\, dx \wedge dy$ is, in fact, holomorphic at every point of S; in other words, its divisor $\Delta(\omega)$ is *positive*, and p_g is precisely the dimension $l(\Delta(\omega))$ of the system of canonical divisors.

But here, almost immediately, an important difference with the theory of curves arose: still following the model of Riemann's theory of "adjoints," it is not difficult to see that for a surface S, defined by an (nonhomogeneous) equation $F(x, y, z) = 0$ of degree n and having no singular point at infinity, the holomorphic differentials $R(x, y, z)\, dx \wedge dy$ are of the form:

$$\frac{Q(x, y, z)}{\dfrac{\partial F}{\partial z}}\, dx \wedge dy,$$

where Q is a polynomial of degree $n - 4$. If S has only ordinary singularities, the surface $Q(x, y, z) = 0$, moreover, must contain the double curve Γ. But it can be verified that if Γ is of degree m, of genus p, and possesses t triple points, the polynomials Q of this form satisfy linear conditions that, if *independent*, would imply that they depend linearly on:

$$(30) \qquad p_a = \binom{n-1}{3} - m(n-4) + 2t + p - 1$$

parameters, the number, then, expected to be equal to p_g. In fact, as early as 1871, Cayley, calculating p_a and p_g for a *ruled* surface, that is, a surface birationally equivalent to the product of a curve of genus p and a line $\mathbf{P}_1(\mathbf{C})$, obtained the values $p_g = 0$ and $p_a = -p$. More generally, for the product surface of two curves of genus π and π', the values $p_g = \pi\pi'$ and $p_a = \pi\pi' - \pi - \pi'$ are obtained. We will see later ((VII, 17) and (VIII, 6)) other interpretations of p_a.

47. When $p_g > 0$, that is to say, when the space $|\Delta(\omega)|$ of positive canonical divisors is nonempty, these divisors are irreducible curves, except for certain particular surfaces. Thus, following M. Noether, the *genus* of the general curve of this linear system, called the *linear genus* of the surface and denoted $p^{(1)}$, can be considered. These three invariants p_g, p_a, and $p^{(1)}$ offer the interest of being *absolute* birational invariants (VI, 32) for surfaces having only ordinary singularities.

From about 1890, the second generation of Italian geometers, among whom Castelnuovo, Enriques, and a little later, Severi stand out above all, undertakes an intensive study of these invariants to which others are added, particularly, the *plurigenera* of Enriques, that are defined by the relations $P_k = l(k \,.\, \Delta)$ where k is an integer $\geqslant 2$ and Δ is a canonical divisor (it is possible that $p_g = l(\Delta)$ is zero, but that the multiples $k \,.\, \Delta$ for $k \geqslant 2$ have equivalent positive divisors). The impulse given to the theory and the harvest of results obtained in a few years are all the more remarkable as the Italian geometers, by working only with positive divisors and most often with surfaces embedded in $\mathbf{P}_3(\mathbf{C})$, impose limitations on themselves that lead to considerable complications in their definitions and techniques. Unfortunately, the very widespread tendency of this school to lack precision in definitions and proofs, facilitated the involvement in numerous futile controversies such as the one that occurred up to the late date of 1943 between Enriques and Severi (see the Bibliography).

Later, we will set forth, in modern language, the results of the Italian school concerning linear systems of curves (the Riemann-Roch problem (VIII, 13), characteristic series (VIII, 17) and (IX, 84)) and their generalizations ("non-linear" systems (VII, 52) and (IX, 84)). However, we mention, forthwith, two problems that have been at the center of the Italian geometers' research.

48. 1° By definition, the number $q = p_g - p_a$ is always $\geqslant 0$ for a surface, and, for nonsingular surfaces, embedded in $\mathbf{P}_3(\mathbf{C})$, $q = 0$; this leads to the name of *irregularity* of the surface for q. Rather quickly, it became apparent that this number is related to the 1-dimensional homology of the surface. Picard had already observed that, in general, a nonsingular surface embedded in $\mathbf{P}_3(\mathbf{C})$ has no simple integrals of the first kind, which is tantamount to saying that its first Betti number $R_1 = 0$; in 1901, Enriques proved, on the contrary, that a surface of *irregularity* > 0 has such integrals. Using "transcendental" results of Picard and a theorem of

Enriques on the "characteristic series" (the proof of which was later found to be incorrect), Severi and Castelnuovo succeeded in proving in 1905 that the numbers q and $\frac{1}{2}R_1$ are equal, and that they are also equal to the maximum number of linearly independent simple integrals of the first kind.

An entirely correct proof of this result (as well as of the theorem of Enriques mentioned above) was obtained for the first time in 1910 by H. Poincaré using a very original, new method. With the notation of (VI, 38), Poincaré considers, on each curve C_y, p abelian integrals of the first kind:

$$I_h(M) = \int_A^M \frac{P_h(x, y, z)\, dx}{\dfrac{\partial F}{\partial z}} \qquad (1 \leqslant h \leqslant p),$$

where P_h is a polynomial in x, y, z and the integral is taken along a path going from A to M on the curve C_y; for each algebraic curve Γ on S, he considers the points of intersection M_k $(1 \leqslant k \leqslant r)$ of C_y and Γ, and associates to Γ, p *normal functions*:

$$v_h(y) = \sum_{k=1}^{r} I_h(M_k) \qquad (1 \leqslant h \leqslant p).$$

If $\Omega_{hj}(y)$ is the period of I_h along the "evanescent cycle" δ_j, Poincaré then shows that, for a suitable choice of polynomials P_h:

$$v_h(y) = \sum_{j=1}^{\mu} \frac{\lambda_j}{2i\pi} \int_a^{y_j} \frac{\Omega_{hj}(u)\, du}{u - y} + c_h \qquad (1 \leqslant h \leqslant q)$$

$$v_h(y) = \sum_{j=1}^{\mu} \frac{\lambda_j}{2i\pi} \int_a^{y_j} \frac{\Omega_{hj}(u)\, du}{u - y} \qquad (q + 1 \leqslant h \leqslant p),$$

where the c_h are constants and the coefficients $\lambda_j = (m \,.\, \delta_j)$ with $m = m_1 + m_2 + \cdots + m_k$, m_k being a path on C_a that deforms, continuously with y, to give a path of integration between A and M_k on C_y. The essential idea is that the systems of functions (v_h) can serve to "parametrize" the algebraic curves Γ traced on S; in particular, Poincaré proves that Γ can be made to vary continuously so that the q constants c_h take *arbitrary* values, and from this, he deduces the existence of a "continuous nonlinear system" (VII, 52) of algebraic curves on S, which was precisely the result needed to complete Severi and Castelnuovo's proof.

49. 2° Whereas algebraic curves have only a single numerical birational invariant, their genus, the existence of numerous invariants of this nature for surfaces renders the problem of classification of algebraic surfaces, from the birational point of view, considerably more difficult. For example, the condition $p_g = p_a = 0$ is not sufficient for a surface to be *rational* (i.e. birationally equivalent to $\mathbf{P}_2(\mathbf{C})$); but Castelnuovo succeeded in obtaining the relations $p_a = 0$ and $P_2 = 0$ as necessary and sufficient condition for rationality. Similarly, the condition $p_a < -1$ implies that the surface is birationally equivalent to a ruled surface, and a necessary and

sufficient condition that this be so is Enriques's criterion: $P_4 = P_6 = 0$ (or $P_{12} = 0$). As a more specific result in the same direction, we mention further that a surface for which $p_a = P_3 = 0$ and $P_2 = 1$ is birationally equivalent to Enriques's surface of the sixth degree that has the six edges of a tetrahedron as double lines (it is not rational although $p_g = 0$).

5. Rational surfaces and Cremona transformations

50. We must back up a little to consider a more specific issue that was essentially initiated by the first generation of Italian geometers and would serve as a model for the more difficult study of linear systems on arbitrary algebraic surfaces: it is the theory of *linear systems of plane curves*, closely linked to the study of *rational surfaces* and *Cremona transformations in the plane*.

A linear system of curves in the projective plane $\mathbf{P}_2(\mathbf{C})$ is defined as being formed of the curves with homogeneous equation:

$$\lambda_0 F_0(x_0, x_1, x_2) + \cdots + \lambda_r F_r(x_0, x_1, x_2) = 0 \qquad (31)$$

where the F_j are linearly independent homogeneous polynomials of the same degree n, and the λ_j are complex parameters. It may always be assumed that the F_j have no common factor of degree > 0, in other words, that the curves (31) have no irreducible component in common. This being so, to each point M of $\mathbf{P}_2(\mathbf{C})$ with homogeneous coordinates x_0, x_1, x_2 can be associated (generalizing (VI, 25)) the point M′ of $\mathbf{P}_r\mathbf{C}$) with homogeneous coordinates $F_j(x_0, x_1, x_2)$, except if all the polynomials are zero at the point M, that is, if M is a point common to the curves (31); these (finitely many) points are called "base points" of the linear system (31). When M tends toward one of the base points O_i along a curve having a tangent at this point, M′ tends toward a limit in $\mathbf{P}_r(\mathbf{C})$, and all these limits are, in general, the points of an algebraic curve Γ_i; the rational transformation $T : M \mapsto M'$ is said to *blow up* the point O_i to the curve Γ_i. It will be assumed that the set $T^{-1}(T(M))$ is, in general, finite. The closure in $\mathbf{P}_r(\mathbf{C})$ of the image under T of the complement in $\mathbf{P}_2(\mathbf{C})$ of the set of base points is also the union of this image and the Γ_i, and is an algebraic surface S that can be proved to be *rational* (a theorem of Castelnuovo, generalizing Lüroth's theorem to dimension 2). T defines a (1, N) correspondence between $\mathbf{P}_2(\mathbf{C})$ and S, where N is the number of points of $T^{-1}(T(M))$ for M in general position; the sets $T^{-1}(T(M))$ are said to form an *involution* of order N on $\mathbf{P}_2(\mathbf{C})$. There exists, in general, a finite number of algebraic curves Θ_k in $\mathbf{P}_2(\mathbf{C})$ such that all the points M of Θ_k (other than the base points) have the same image in S; these are called the *fundamental curves of T* (or of the system (31)). The images under T of the curves (31) are the *sections* of S by the *hyperplanes* of $\mathbf{P}_r(\mathbf{C})$. Finally, we note that a theorem of Bertini shows that the curves of the system (31) can not have singular points varying with the λ_j; in other words, except for isolated values of the λ_j, the singular points of the curves of the system are all situated at the base points.

51. The most interesting case is the case where $N = 1$, that is, where the curves (31) passing through M (M, a point distinct from the base points and not on a fundamental curve) have no point in common other than M and the base points. In this case, T^{-1} is called a "plane representation" of S. As T is birational, the sections of S by the hyperplanes of $\mathbf{P}_r(\mathbf{C})$ have the same genus as the curves (31). The investigation of the algebraic curves contained in S is greatly facilitated by the existence of a plane representation, which brings this study back to that of plane curves.

For example, if S is a nondegenerate quadric in $\mathbf{P}_3(\mathbf{C})$ and T^{-1} is the projection of S on a plane from a point A of S, T is obtained by taking for (31) the family of conics passing through two distinct base points O_1, O_2 of $\mathbf{P}_2(\mathbf{C})$, which "blow up" to the two generators D_1, D_2 of S passing through A; there is a single fundamental curve, the line O_1O_2, to which the point A in S corresponds. An algebraic curve C′ on S not passing through A has two "degrees" m_1, m_2, which are the intersection numbers of C′ with generators of the same system as D_1 and D_2, respectively. These generators correspond under T to lines passing through O_2 and O_1, respectively, and C′ is the image under T of a curve C of degree $m_1 + m_2$ having O_1 a multiple point of order m_1 and O_2 a multiple point of order m_2. Note that C has no other point in common with O_1O_2. It can easily be shown that if $m_1 = m_2 = m$, then C′ is the complete intersection of S and a surface of degree m; if, on the contrary, $m_1 > m_2$, for example, then there exists a surface of degree m_1 that cuts out C′ and $m_1 - m_2$ generators of the system of D_2 from S.

52. Now, for the linear system (31), take the system of cubics passing through 6 base points O_i $(1 \leqslant i \leqslant 6)$. If these points are not on the same conic, the system defines a birational transformation T of $\mathbf{P}_2(\mathbf{C})$ onto a cubic surface S of $\mathbf{P}_3(\mathbf{C})$ and, conversely, a general cubic surface admits a plane representation T^{-1} obtained in this manner. There is no fundamental curve; the points O_i blow up to 6 lines on S; 21 other lines on S are obtained by considering the images of the 15 lines O_iO_j for $i \neq j$ and the images of the 6 conics passing through 5 of the points O_i. The properties of these lines follow very easily from this definition. If the points O_i are on the same nondegenerate conic, this conic is the unique fundamental curve to which corresponds, this time, a double point on the cubic surface S, and there are no more than 21 lines on S. Similarly, cubic surfaces having 2, 3, or 4 double points are obtained by taking the O_i on degenerate conics; for example, if the O_i are the vertices of a complete quadrilateral, a cubic surface with 4 double points, containing only 9 lines, is obtained.

53. More generally, if instead of taking the cubics passing through 6 points O_i, the cubics passing through $9 - n$ points with $3 \leqslant n \leqslant 9$ are taken for the system (31), a birational transformation of $\mathbf{P}_2(\mathbf{C})$ onto a surface S of degree n in a space $\mathbf{P}_n(\mathbf{C})$ of dimension *equal* to the degree of the surface is obtained. But, in contrast to the fact that there are nonsingular rational curves of *arbitrary* degree n in a space $\mathbf{P}_n(\mathbf{C})$ (VI, 29), the preceding surfaces, as well as the surface of degree 8 in $\mathbf{P}_8(\mathbf{C})$

obtained by taking the system of quartics having 2 fixed double points for the system (31), are the *only* nonsingular rational surfaces of degree n contained in $\mathbf{P}_n(\mathbf{C})$ and not in a hyperplane of $\mathbf{P}_n(\mathbf{C})$; these are called *Del Pezzo surfaces*. The Del Pezzo surface of degree 4 (also called the *Segre surface*) is the intersection of 2 quadrics (of dimension 3) in $\mathbf{P}_4(\mathbf{C})$.

Finally, if the system (31) of the cubics passing through 7 base points O_i is considered, a birational transformation is no longer obtained, but actually, a (1, 2) transformation of $\mathbf{P}_2(\mathbf{C})$ on itself, as all the cubics passing through 8 points have necessarily a ninth point in common. The corresponding involution of order 2 in $\mathbf{P}_2(\mathbf{C})$ is called *Geiser's involution* (cf. (VI, 57)).

54. If, for the system (31), all the conics in the plane are taken (which is the same as taking all the monomials of the second degree for the F_j), a birational transformation T of $\mathbf{P}_2(\mathbf{C})$ onto a surface V of degree 4 in $\mathbf{P}_5(\mathbf{C})$, called the *Veronese surface*, is obtained; there are neither base points nor fundamental curves and T is bijective. The surface V has remarkable properties: the irreducible curves that it contains are all of even degree, and there is, on V, a two-parameter family of conics corresponding to the lines of $\mathbf{P}_2(\mathbf{C})$. By projecting V into $\mathbf{P}_3(\mathbf{C})$ from a line not meeting V in $\mathbf{P}_5(\mathbf{C})$, a *Steiner surface* is obtained. It is a surface of fourth degree having 3 double lines meeting at a triple point, with 2 pinch-points found on each of them; the tangent planes to this surface cut out 2 conics from it. Conversely, it can be shown that the only surfaces in $\mathbf{P}_3(\mathbf{C})$ whose sections by their tangent planes are *reducible* curves are the Steiner surface and the ruled surfaces (theorem of Kronecker-Castelnuovo).

Lastly, the "plane representation" method can be applied to *rational ruled surfaces*: it suffices to take for (31) the system of curves of degree n having n base points, one of which is of order $n - 1$ and the $n - 1$ others are simple. A birational transformation T of $\mathbf{P}_2(\mathbf{C})$ onto a ruled surface S of degree n in $\mathbf{P}_{n+1}(\mathbf{C})$ is obtained with generators corresponding to the lines of $\mathbf{P}_2(\mathbf{C})$ passing through the multiple point of order $n - 1$. It can be shown that these surfaces are, with the Veronese surface V, the only surfaces of degree n in $\mathbf{P}_{n+1}(\mathbf{C})$ not contained in a hyperplane of $\mathbf{P}_{n+1}(\mathbf{C})$.

55. Cremona also connected the determination of all the birational transformations of $\mathbf{P}_2(\mathbf{C})$ onto itself (called *Cremona transformations* of the plane) to the preceding method: indeed, it suffices to take the number $r = 2$ in (31), the curves of the system being additionally constrained by the condition that any two of them have only a single point in common other than the base points (such a system is called a *homaloidal net*). This system defines a birational transformation T of $\mathbf{P}_2(\mathbf{C})$ on itself, which carries the curves of the net into the lines of the plane. Conversely, for every birational transformation T of the plane, the transforms under T^{-1} of the lines of the plane form a homaloidal net that defines T. It is clear that the curves of the net must be *rational*, which implies the relation:

$$(32) \qquad (n-1)(n-2) - \sum_i k_i(k_i - 1) = 0$$

between their degree n and their multiplicities k_i at the base points O_i of the net, and, from the fact that two curves of the net have only a single point in common besides the base points, the relation:

$$(33) \qquad n^2 - \sum_i k_i^2 = 1$$

is obtained, two conditions, to which must be added that the curves of the net are irreducible. For small values of n, it is easy to find all the corresponding Cremona transformations by examining solutions of the diophantine system formed by (32) and (33).

56. As examples of Cremona transformations, there is, of course, the quadratic transformation (IV, 11), corresponding to $n = 2$, with 3 simple base points. We point out also, for every integer n, the *de Jonquières transformation* where there is a multiple base point O of order $n-1$ and $2n-2$ simple base points A_i; the fundamental curves are the $2n-2$ lines OA_i and a curve of degree $n-1$ passing through the A_i and having O as a multiple point of order $n-2$; the inverse transformation is also a de Jonquières transformation.

A fundamental theorem from the theory of Cremona transformations in the plane is that of Noether-Castelnuovo; every Cremona transformation in $\mathbf{P}_2(\mathbf{C})$ is a product of a finite number of quadratic transformations.

57. The involutions on $\mathbf{P}_2(\mathbf{C})$ (VI, 50) were also thoroughly studied; in particular, an involution of order 2 evidently defines a Cremona transformation equal to its inverse. It turns out that the latter are known (theorem of Bertini); they are (up to a birational transformation):

1) the involutive projective transformation (in nonhomogeneous coordinates, symmetry with respect to a line);

2) a de Jonquières involution: given a curve Γ of degree n with multiple point O of order $n-2$, to a point M is associated its harmonic conjugate with respect to the two points, other than O, where OM meets the curve Γ;

3) Geiser's involution (VI, 53);

4) Bertini's involution: starting with the system of the sextics having 8 given double points, for each point M in $\mathbf{P}_2(\mathbf{C})$, the cubic passing through M and the O_i meets all the sextics of the family passing through M in a second fixed point M′ (otherwise the cubic would be rational), and it is M′ that corresponds to M under Bertini's involution.

More generally, all the finite groups of Cremona transformations have been determined, as well as all the Lie groups formed of such transformations (Enriques).

58. Linear systems of hypersurfaces in $\mathbf{P}_n(\mathbf{C})$, the rational varieties of dimension n that they define, and Cremona transformations in $\mathbf{P}_n(\mathbf{C})$ can be treated in the same way, but the results are much less complete, and, in particular, there is no analogue of the Noether-Castelnuovo theorem.

Another generalization consists in replacing $\mathbf{P}_2(\mathbf{C})$ by an arbitrary algebraic surface, which is an aspect of the theory of linear systems brought out above (VI, 45). Thus, the notion of exceptional curve on an algebraic surface arises; it is, by definition, a fundamental curve (in the sense of (VI, 50)) for a birational transformation of one surface into another (see (IX, 18)).

VII — THE SIXTH EPOCH

New Structures in Algebraic Geometry (1920–1950)

At the beginning of the twentieth century the general idea of the *structures* underlying diverse mathematical theories became completely conscious and, as a consequence, led to a remodeling and deepening of most of these theories, underscoring more and more the deep unity of mathematics. Algebraic geometry is one of the principal beneficiaries of these tendencies; its relations with the theory of analytic functions, on the one hand, and with algebra and number theory, on the other, are expanded considerably, enriching it with a multitude of new points of view. One of the effects of this expansion will be to erode, little by little, the rigid framework of the projective and birational methods, and to prepare the way for a much more flexible and more general concept.

1. Kähler manifolds and return to Riemann

1. The intrinsic concept of *differential manifold*, independent of any embedding in a numerical space, dates back to Gauss's 1826 memoir on the theory of surfaces, and was expressed for the first time in its general form by Riemann in his inaugural lecture of 1854. At the beginning of the twentieth century, this concept had acquired a precise form: the notion of a differential manifold of dimension n consists of a topological space M equipped with "charts," homomorphisms $\phi : U \to \mathbf{R}^n$ from open sets U of M onto open sets of $\mathbf{R}^n$ such that the domains of the charts cover M, and, for any two charts $\phi : U \to \mathbf{R}^n$, $\psi : V \to \mathbf{R}^n$, the "transition function" $z \mapsto \psi(\phi^{-1}(z))$ is a diffeomorphism of the open set $\phi(U \cap V)$ onto the open set $\psi(U \cap V)$; then, the coordinates $u_1, \ldots, u_n$ of the point $\phi(x)$ of $\phi(u)$ are called the local coordinates of x for the chart ϕ.

2. One of the most important developments of this theory was the introduction, by H. Poincaré and E. Cartan, of the concept of *exterior differential p-form* on a differential manifold M; in the domain of a chart of M, one such form can be expressed, using local coordinates, by:

$$(1) \qquad \omega = \sum_{i_1 < i_2 < \cdots < i_p} A_{i_1 i_2 \cdots i_p}(u_1, \ldots, u_n)\, du_{i_1} \wedge du_{i_2} \wedge \cdots \wedge du_{i_p}$$

where the functions (with complex values) $A_{i_1 i_2 \cdots i_p}$ are of class C^∞. Recall that for these forms, on the one hand, the *exterior differential* of a p-form ω can be defined as the $(p+1)$-form that, for the local expression (1) of ω, is given by:

$$(2) \qquad d\omega = \sum_{i_1 < i_2 < \cdots < i_p} dA_{i_1 i_2 \cdots i_p} \wedge du_{i_1} \wedge du_{i_2} \wedge \cdots \wedge du_{i_p},$$

and satisfies the fundamental relations:

$$(3) \qquad d(d\omega) = 0$$

$$(4) \qquad d(\omega_p \wedge \omega_q) = (d\omega_p) \wedge \omega_q + (-1)^p \omega_p \wedge (d\omega_q)$$

for a p-form ω_p and a q-form ω_q. On the other hand, generalizing the classical process of integration on a curve or a surface, the integral of a p-form ω on an oriented singular p-simplex (cf. (VI, 30)) can be defined, and then, by linearity, the integral $\int_C \omega = \langle C, \omega \rangle$ of ω on an arbitrary p-chain. These notions are connected by the generalized Stokes formula:

$$(5) \qquad \langle C, d\omega \rangle = \langle \partial C, \omega \rangle$$

for a differential p-form ω and a $(p+1)$-chain C.

3. As early as the beginning of his work on algebraic topology, H. Poincaré had explored the relations between the homology of a compact differential manifold M and the exterior differential forms on M. These relations were put in a precise form in de Rham's celebrated theorem of 1931: let Λ^p be the vector space of *real* differential p-forms on M, and Λ the "complex"

$$(6) \qquad 0 \rightarrow \Lambda^1 \xrightarrow{d} \Lambda^2 \xrightarrow{d} \cdots \rightarrow \Lambda^n \rightarrow 0 \quad (n = \dim M)$$

of vector spaces over **R**; the relation (3) allows "cohomology spaces" $H^i(\Lambda)$ $(1 \leqslant i \leqslant n)$ to be associated to this complex, the space $H^i(\Lambda)$ being defined as the quotient of the kernel of $d: \Lambda^i \rightarrow \Lambda^{i+1}$ (the space of *closed* i-forms) by the image of $d: \Lambda^{i-1} \rightarrow \Lambda^i$. The theorem of de Rahm establishes that the duality (5) between p-forms and p-chains carries over to a duality between the homology spaces $H_i(M, \mathbf{R})$ and the cohomology spaces $H^i(\Lambda)$ (denoted also by $H^i(M, \mathbf{R})$).

Now we recall that in the cohomology space $H^i(M, \mathbf{R})$, one singles out the subgroup $H^i(M, \mathbf{Z})$ (cohomology with *integer* values) of the classes of closed forms taking integer values on the *simplices* of the same dimension, and the **Q**-vector space $H^i(M, \mathbf{Q})$ (cohomology with *rational* values) of classes of closed forms with rational values on the simplices of the same dimension. By definition, $H^i(M, \mathbf{C}) = H^i(M, \mathbf{R}) \otimes_{\mathbf{R}} \mathbf{C}$.

Also, the direct sum $H^\bullet(M, \mathbf{R})$ (resp. $H^\bullet(M, \mathbf{C})$, $H^\bullet(M, \mathbf{Z})$, $H^\bullet(M, \mathbf{Q})$) of the $H^i(M, \mathbf{R})$ (resp. $H^i(M, \mathbf{C})$, $H^i(M, \mathbf{Z})$, $H^i(M, \mathbf{Q})$) is considered with the structure of a noncommutative *ring*, called the *cohomology ring*, canonically defined on it; it is an algebra over **R** (resp. **C**, **Z**, **Q**), the product being induced by passage to the quotient of the exterior product of the differential forms.

4. From 1931 on, in a series of remarkable memoirs, Hodge, generalizing Riemann's methods, showed how to obtain another extremely useful interpretation of the homology of a compact differential manifold and how to apply it to

algebraic varieties. On a differential manifold M, there can be defined (in an infinite number of ways) a riemannian metric $ds^2 = \sum_{i,j} g_{ij}(u_1, \ldots, u_n)\, du_i du_j$ (in local coordinates); Beltrami had showed that an operator (dependent on ds^2) that generalizes the usual laplacian can be defined on a Riemann manifold, and, consequently, that the notion of *harmonic function* can be defined on such a manifold. Hodge's originality lay in seeing, more generally, that the notion of *harmonic exterior differential form* could be defined on a Riemann manifold M and that it was here that the extension to arbitrary dimensions of the theory of abelian integrals on a Riemann surface must be found. At each point of M, the riemannian metric defines a nondegenerate, positive, symmetric bilinear form on the tangent vectors, and, consequently, also on all the tensor products, by the usual process of multilinear algebra. Therefore, on the space of real differential p-forms on M, there is, canonically defined from the riemannian structure, a nondegenerate symmetric bilinear form $(\alpha, \beta) \mapsto \mathbf{g}_p(\alpha, \beta)$. If M is orientable and of dimension n, this allows, canonically, the association of a $(n - p)$-form denoted $*\alpha$ to each p-form α, characterized by the relation:

$$(7) \qquad \beta \wedge (*\alpha) = \mathbf{g}_p(\alpha, \beta) v$$

for all differential p-forms β, where v is the canonical "volume" n-form on M. The duality thus established between p-forms and $(n - p)$-forms permits the definition, from the operator d, of a new differential operator $\delta = -(*) \circ d \circ (*)$ that transforms a p-form into a $(p - 1)$-form, and then the definition of an operator $\Delta = d \circ \delta + \delta \circ d$, transforming p-forms to p-forms, which is the generalization of Beltrami's laplacian (to which it reduces (up to sign) for $p = 0$). A *harmonic p-form* α is defined by the condition $\Delta\alpha = 0$, and Hodge's fundamental theorem is that if M is a compact, orientable Riemann manifold, there exists a unique harmonic p-form in each cohomology class of $H^p(\Lambda)$, and, consequently, the space $H^p(\Lambda)$ is isomorphic to the vector space $\mathbf{H}^p$ of harmonic p-forms on M. In particular, it follows that there exists a unique harmonic p-form α having fixed *periods* $\langle C_j, \alpha \rangle$ for R_p p-cycles C_j, whose homology classes form a base of $H_p(M, \mathbf{R})$. This is the generalization of Riemann's fundamental result and shows that Riemann's use of the "Dirichlet principle" was certainly not just a technical artifice (happily for Hodge, the theory of elliptic partial differential equations was advanced enough in 1930 to provide the means for a rigorous proof that Riemann had lacked).

Note that if α is a harmonic p-form, $*\alpha$ is a harmonic $(n - p)$-form, thus recovering the Poincaré duality $R_{n-p} = R_p$.

5. The possibility of applying Hodge's theory to nonsingular projective complex varieties results, first of all, from the fact that such a variety M of (complex) dimension n has a natural structure of a compact, orientable differential manifold of dimension $2n$. But, in fact, it has an even richer structure. In the first place, it is a *holomorphic* manifold, which means that there are charts in $\mathbf{R}^{2n}$ (identified with $\mathbf{C}^n$) whose domains of definition cover M, and for which the "transition functions" (VII, 1) are *holomorphic*. For an arbitrary holomorphic manifold M of (complex) dimension n (therefore, of real dimension $2n$), there is a canonical decomposition of

the complex differential p-forms, corresponding to the pairs of integers r, s such that $r \leqslant n, s \leqslant n$ and $r + s = p$: a differential p-form is said to be *of type* (r, s) if, for local *complex* coordinates $z_1, \ldots, z_n$, it can be written:

$$\sum_{j_1, \ldots, j_r, k_1, \ldots, k_s} A_{j_1 \cdots j_r k_1 \cdots k_s}(z_1, \ldots, z_n)\, dz_{j_1} \wedge \ldots \wedge dz_{j_r} \wedge d\bar{z}_{k_1} \wedge \ldots \wedge d\bar{z}_{k_s}, \tag{8}$$

where the $A_{j_1 \cdots k_s}$ are functions of class C^∞ with complex values (not, in general, holomorphic). This definition is independent of the choice of local complex coordinates, and every p-form can be written uniquely in the form:

$$\omega = \sum_{r+s=p} \omega^{r,s}, \tag{9}$$

where $\omega^{r,s}$ is of type (r, s) (with $\omega^{r,s} = 0$ if $r > n$ or $s > n$).

6. On each holomorphic manifold of complex dimension n, *hermitian* ds^2 can be defined, that is, such a ds^2 can be expressed, using local complex coordinates $z_1, \ldots, z_n$, as a hermitian form in the dz_j:

$$ds^2 = \sum_{j,k} h_{jk}\, dz_j\, d\bar{z}_k \qquad \text{with} \qquad h_{kj} = \bar{h}_{jk}. \tag{10}$$

From linear algebra, it is known that the imaginary part of every hermitian form is an *alternating* form; thus to a hermitian ds^2 on a holomorphic variety M corresponds canonically its imaginary part that, therefore, can be written as a *real* exterior differential 2-*form*:

$$\Theta = \frac{i}{2} \sum_{j,k} h_{jk}\, d\bar{z}_k \wedge dz_j. \tag{11}$$

In 1933, E. Kähler observed that a hermitian metric ds^2 can be defined on complex projective space $\mathbf{P}_n(\mathbf{C})$ for which the form Θ is closed, that is, $d\Theta = 0$. As Hodge showed, this property implies a whole series of remarkable consequences for harmonic differential forms, whence the name *Kähler manifold* given to holomorphic manifolds equipped with a kählerian metric (that is, such that $d\Theta = 0$).

In the sequel, the operator Δ is canonically extended to *complex* exterior differential forms, yielding the notion of complex harmonic form; now, we will denote by $\mathbf{H}^p$ the space of complex harmonic p-forms (obtained by extension of scalars from the space of real harmonic p-forms); therefore, it is isomorphic to $H^p(M, \mathbf{C})$.

7. In the first place, Δ, then, commutes with the projection operators $\omega \mapsto \omega^{r,s}$ from the space of p-forms onto that of the forms of type (r, s) $(r + s = p)$; thus, if ω is harmonic, so is $\omega^{r,s}$, and

$$\mathbf{H}^p = \bigoplus_{r+s=p} \mathbf{H}^{r,s}$$

where $\mathbf{H}^{r,s}$ designates the space of harmonic forms of type (r, s). Therefore, for a *compact* Kähler manifold, from this remark and from the Hodge-de Rham theorem (VII, 4), it follows that if $h^{r,s}$ is the (finite) dimension of $\mathbf{H}^{r,s}$, then the pth Betti number satisfies the formula:

(12) $$R_p = \sum_{r+s=p} h^{r,s} \qquad (1 \leqslant p \leqslant 2n).$$

Moreover, it follows from the definition (8) that if ω is of type (r, s), its complex conjugate $\bar{\omega}$ is of type (s, r) and, therefore:

(13) $$h^{s,r} = h^{r,s}.$$

From this and from (12) it follows that if p is *odd*, R_p is *even*.

In the second place, we denote by Θ^k the $2k$-form that is the exterior product of k 2-forms identical to Θ; it is easily shown that the Θ^k $(1 \leqslant k \leqslant n)$ are all harmonic and $\neq 0$, and that Θ^n is, up to nonzero constant factor, the "volume" form on M. Therefore, it is seen that $\mathbf{H}^{2k}$ is not reduced to 0 for $1 \leqslant k \leqslant n$, in other words, $R_{2k} \geqslant 1$.

8.   Consider the linear map $L : \omega \mapsto \Theta \wedge \omega$; on a Kähler manifold, it can be shown that L commutes with $\varDelta$ and, consequently, that L transforms a harmonic form into a harmonic form. A harmonic p-form ω is said to be *primitive* if $p \leqslant n$ and if $L^{n-p+1}\omega = 0$. $\mathbf{P}^{r,s}$ denotes the subspace of $\mathbf{H}^{r,s}$ composed of the primitive harmonic forms for $r + s \leqslant n$, and $\rho^{r,s}$, its dimension. If $r + s < n$, it can be shown that $\mathbf{H}^{r+1,s+1}$ is the direct sum of $L(\mathbf{H}^{r,s})$ and of $\mathbf{P}^{r+1,s+1}$, the map L being injective in $\mathbf{H}^{r,s}$, thus giving the formula:

(14) $$h^{r+1,s+1} = h^{r,s} + \rho^{r+1,s+1} \qquad (r + s < n)$$

from which follows:

(15) $$R_p - R_{p-2} = \sum_{r+s=p} \rho^{r,s} \geqslant 0 \qquad \text{for} \qquad p \leqslant n;$$

moreover, the same argument as for (13) shows that

(16) $$\rho^{s,r} = \rho^{r,s}.$$

In addition, Hodge proved that the numbers $\rho^{r,s}$ are independent of the kählerian metric chosen on M.

We note also the important fact that the operator

(17) $$L^{n-p} : \mathbf{H}^p \to \mathbf{H}^{2n-p} \qquad (\text{for } p < n)$$

is an *isomorphism*, making Poincaré duality considerably more precise.

We assume that the (complex) dimension n of M is *even*. Then, a symmetric bilinear form is defined on $\mathbf{H}^n : (\alpha, \beta) \mapsto \int \alpha \wedge \beta$, which is real when α and β are; by restricting to real n-forms, it is shown that this bilinear form is nondegenerate and that its signature is given by

$$\left(\sum_{\substack{r,s \text{ even} \\ r+s \leqslant n}} \rho^{r,s}, \sum_{\substack{r,s \text{ odd} \\ r+s \leqslant n}} \rho^{r,s} \right)$$

(Hodge's index theorem)

9.   On a holomorphic manifold, the *holomorphic* differential p-forms (also said to be *of the first kind*) are, by definition, those of type $(p, 0)$ for which the expression (8) using complex local coordinates $z_1, \ldots, z_n$ has holomorphic coefficients (a

condition that is independent of the local coordinates chosen). On an arbitrary holomorphic manifold, a closed form of type $(p, 0)$ is holomorphic, and if the manifold is Kähler, a holomorphic form of type $(p, 0)$ is (complex) harmonic; however, in general, the converses of these properties are false. Nevertheless, on a *compact Kähler* manifold for closed forms of type $(p, 0)$, the closed forms, the holomorphic forms, and the (complex) harmonic forms are *identical*. From this, it follows that a holomorphic p-form with zero for all its periods is necessarily zero, because, in this case, it is of the form $d\omega$ and the only forms of this type that are of type $(p, 0)$ come from holomorphic forms of type $(p-1, 0)$.

10. The metric induced on each closed complex submanifold of a Kähler manifold is again Kähler. In particular, an *algebraic* projective variety can be equipped with a kählerian metric, and, therefore, the results from Hodge theory are applicable; thus, specifically, the properties of the Betti numbers discovered by Lefschetz (VI, 38) are recovered. However, to be able to apply, for example, the results on holomorphic p-forms on a compact Kähler manifold to the properties of *rational* p-forms on a nonsingular projective algebraic variety V (that is to say, the forms whose expression, using local coordinates belonging to the field of rational functions on V, has coefficients also belonging to this field [cf. (VI, 44)]), it is necessary to know that a holomorphic p-form is necessarily rational. This fact, which follows from Riemann's theory for algebraic curves, appears to have been admitted without proof until Serre's work, in 1956, which treated this question thoroughly in the more general framework of the theory of coherent sheaves (see (VIII, 24)). As an example of the application of Hodge theory, the decomposition $\mathbf{H}^1 = \mathbf{H}^{1,0} \oplus \mathbf{H}^{0,1}$ and the equality (13) give, for every projective algebraic variety V, the equality $\frac{1}{2}R_1 = h^{1,0}$ (the number of simple integrals of the first kind) which, for surfaces, is part of the fundamental result of (VI, 48) (the equality $q = \frac{1}{2}R_1$ resulting, as will be seen, from the cohomology of sheaves). Another example is a result announced by Lefschetz with incomplete proof: for a nonsingular algebraic surface S, an isomorphism $H_3(S, \mathbf{Q}) \xrightarrow{\sim} H_1(S, \mathbf{Q})$ is defined by mapping the class of a 3-cycle on S to the class of its intersection with a section of S by a hyperplane (which is a 2-cycle); this statement can be shown to follow, by duality, from the isomorphism (17) in the case $n = 2$, $p = 1$.

Thus, Hodge theory places the "transcendental" theory of projective algebraic varieties in a more comprehensive framework, as it is easy to give examples of compact Kähler manifolds not isomorphic to a projective algebraic variety, for example, a *torus* $\mathbf{C}^n/\Gamma$, where Γ is a "general" lattice of $\mathbf{R}^{2n}$ (cf. VII, 55)); the problem of characterizing the projective algebraic varieties among the compact Kähler manifolds was resolved by Kodaira only in 1954 (see (VIII, 15)).

2. Abstract Algebraic Geometry and Commutative Algebra

11. The development of "abstract" algebra begins around 1900 when it is recognized that the notion of *algebraic structure* (such as the structure of group,

ring, field, module, etc.) is the fundamental notion in algebra, putting the nature of the mathematical objects on which such a structure is defined in the background, whereas, up to then, the majority of algebraic theories dealt with calculations principally over the real or complex numbers. Thus it was natural to think of extending the concepts of algebraic geometry to the case where the coefficients of the equations and the coordinates of the points no longer are complex numbers but are elements of an arbitrary (commutative) field.

12. In fact, as early as 1882, Dedekind and Weber had observed, in their memoir on algebraic curves, that their arguments did not use at all the topological properties of the field of complex numbers but only the fact that this field is algebraically closed. To tell the truth, they use, as well, the fact that the field is of characteristic 0 (a notion that had not been defined at this time); but, as F. K. Schmidt showed in 1929, only a very few modifications are necessary to make their results (including the Riemann-Roch theorem) applicable when the base field is an algebraically closed field of arbitrary characteristic.

13. Still, the algebraic techniques of Dedekind and Weber, modeled on those from algebraic number theory, are strictly limited to dimension 1. The algebraic tools necessary to eliminate the intervention of analysis in the general problems of algebraic geometry will be forged, progressively between 1890 and 1950, in a nonsystematic way, and often by algebraists whose preoccupations concern algebraic geometry in only slight measure.

A first, preliminary, task consisted in disentangling the essential notions of the *theory of commutative fields*. Although, even earlier than 1900, there was already a rather wide range of diverse examples of fields, the general theory was founded by Steinitz only in 1910. Among the fundamental concepts and results of his work, one should point out the definition of transcendence degree (needed to define, in general, the notion of *dimension* of an algebraic variety on the model of Kronecker's definition), the existence of an algebraic closure for any field, and the concept of separable algebraic extension (extended to transcendental extensions, around 1939, by S. MacLane), indispensable to the development of algebraic geometry over a field of characteristic $p > 0$.

14. Up to the introduction of "abstract" algebraic varieties by A. Weil in 1946, algebraic varieties are always considered as subsets of an "affine space" k^N or of a "projective space" $\mathbf{P}_N(k)$, defined as the set of points whose coordinates satisfy arbitrary polynomial equations, in the first case, and whose homogeneous coordinates satisfy homogeneous polynomial equations, in the second. In fact, from the beginnings of the algebraic school, Kronecker, like Dedekind and Weber, found himself impeded by the taboo to work only in projective spaces, and was aware that for most of the questions of projective geometry, it sufficed first to treat the problems in affine space, and then to cover projective space by a finite number of affine open sets. From the algebraic point of view, the interest of geometry in

affine space is that algebraic varieties in such a space can be associated to *intrinsically algebraic* objects, the ideals of the polynomial ring $k[T_1, \ldots, T_N]$, whereas for projective varieties, it is necessary to consider *graded* ideals of the polynomial ring $k[T_0, T_1, \ldots, T_N]$, which are not as convenient to handle.

15. In fact, it can be said that up to about 1925, algebraic geometry over an arbitrary field will be restricted to the theory of ideals in polynomial rings and their generalizations, the purely algebraic aspects of this theory greatly outweighing their geometric interpretations. The central problem is that of the *decomposition of ideals* motivated, on the one hand, by the theorem of unique factorization of prime ideals in Dedekind rings, and, on the other, by the decomposition of an algebraic variety into irreducible varieties. The latter was obtained by Kronecker and, yet again, by Lasker using a method of elimination. The progress will consist in substituting arguments from the theory of ideals, inspired by those of Dedekind, owing to the introduction of *primary* ideals by Lasker, and of rings satisfying the maximal condition for ideals (now called noetherian rings) by E. Noether, which leads to the definitive version of the theorem of decomposition into primary ideals in 1921.

16. The initial impulse of the work of Lasker and of E. Noether is to be found in the first general theorems on rings of polynomials in any number of indeterminates, discovered by Hilbert in 1890–1893, occasioned by his celebrated memoirs on invariant theory: the "finite basis" theorem and the "theorem of zeros." Both answer questions derived from the theory of algebraic curves; Cayley and Kronecker had posed the problem of defining each irreducible algebraic space curve as the intersection of a finite number of algebraic surfaces, and Hilbert's theorem showing that each ideal of $\mathbf{C}[T_1, \ldots, T_n]$ (or of $\mathbf{Z}[T_1, \ldots, T_n]$) admits a finite system of generators answered the same question for every affine algebraic variety. On the other hand, the Brill-Noether theory of linear series on a plane curve (VI, 23) rests on a fundamental theorem of M. Noether giving conditions, in $\mathbf{C}[T_1, T_2]$, for a polynomial F to belong to the ideal generated by two given polynomials P, Q (*it is not sufficient* that the curve $F(x,y) = 0$ pass through the points of intersection of the curves $P(x,y) = 0$ and $Q(x,y) = 0$ when these are multiple). A little later, Netto had proved that, without any additional conditions, every polynomial $F(x, y)$ that is zero at all the points common to the two curves $P(x,y) = 0$, $Q(x,y) = 0$ has a *power* F^k belonging to the ideal generated by P and Q. One of the forms of Hilbert's theorem of zeros is that, in general, every polynomial of $\mathbf{C}[T_1, \ldots, T_n]$ that is zero at all the points of $\mathbf{C}^n$ where the polynomials of an ideal $\mathfrak{a}$ of $\mathbf{C}[T_1, \ldots, T_n]$ are zero, has a power belonging to $\mathfrak{a}$.

It is this result that must have led Lasker to define the notion of primary ideal in a polynomial ring by the property that if a product xy belongs to such an ideal $\mathfrak{q}$, and if $y \notin \mathfrak{q}$, then $x^k \in \mathfrak{q}$ for a sufficiently large exponent k.

17. This is a suitable place to mention another of Hilbert's ideas, connected to the finite basis theorem, that was to acquire great importance. It concerns graded

ideals $\mathfrak{a}$ of $\mathbf{C}[T_0, \ldots, T_n]$. Hilbert considers, for each degree m, the dimension of the vector space quotient (over $\mathbf{C}$) of the vector space S_m of homogeneous polynomials of degree m by the subspace $S_m \cap \mathfrak{a}$. He proves that this integer $\chi_{\mathfrak{a}}(m)$ is, *for sufficiently large m*, equal to a *polynomial in m*, a result which, moreover, implies the finite basis theorem.

Then, for m large, there is an expression:

$$\chi_{\mathfrak{a}}(m) = a_0\binom{m}{r} + a_1\binom{m}{r-1} + \cdots + a_r,$$

where the a_j are integers. If $\mathfrak{a}$ is the ideal of a projective variety V in $\mathbf{P}_n(\mathbf{C})$, r is the dimension of V (i.e. the largest dimension of its irreducible components), and $(-1)^r(a_r - 1)$ has been defined by Severi as the *arithmetic genus* of V (it coincides with p_a (VI, 46) in the case of surfaces in $\mathbf{P}_3(\mathbf{C})$).

18. The work of E. Noether and of W. Krull on local rings and the general notion of element integral over a ring begins around 1925; we will discuss it a little later when these purely algebraic notions find their interpretation in the language of algebraic varieties, in the period 1940–1950.

3. Specializations, generic points, and intersections

19. Since Poncelet's time, it has been customary in algebraic geometry to restrict the proofs of most of the general theorems to the case where the points or the algebraic varieties under consideration are "in general position," as the reader will have been able to verify in numerous earlier statements. Most often, the authors neglect to say exactly what they mean by that, relying on the text to clarify their intent; the general idea is that when the points or varieties under consideration depend on a certain number of (complex) variable parameters, they are "in general position" when these parameters do not satisfy a certain (finite) number of *polynomial relations*, depending on the question studied. Stated a little more precisely, if the point of affine or projective space having these parameters for coordinates is considered, the problem studied implies that this point belongs to an algebraic variety V, and the points of V for which the data are not "in general position" are the points of algebraic subvarieties W_j of V, distinct from V. If V is irreducible, it follows that the complement of the W_j is an *open, everywhere dense* subset of V.

20. Poncelet himself had indicated how such ideas could furnish proofs of theorems where no "general position" condition was imposed, for example, Bezout's theorem on the intersection of a curve C of degree m and a curve C′ of degree n (irreducible or not) in $\mathbf{P}_2(\mathbf{C})$. His method consists in considering the set F_n of *all* the (not necessarily irreducible) curves of degree n, which is naturally identified with a projective space $\mathbf{P}_N(\mathbf{C})$ (IV, 9); the conditions for a curve $\Gamma \in F_n$

to have a multiple point belonging to C or to pass through a multiple point of C or to be tangent to C at a simple point are expressed by algebraic relations, and thus define subvarieties W_j in $\mathbf{P}_N(\mathbf{C})$ distinct from the whole space. If Γ is not in one of these varieties, then C and Γ have in common only simple points where their tangents are distinct, and Poncelet accepts as true that the number of these points is constant (a fact that can easily be proved using the continuity of the roots of an equation as a function of the parameters, and the fact that the complement G of the union of the W_j is connected); this number is, moreover, equal to mn, as can be seen by taking a curve composed of n distinct lines for Γ. That being so, Poncelet proposes to take as multiplicity of the intersection of C and C′ at the point P, the number of the points of intersection of C and Γ that tend toward P as Γ tends toward C′ (the existence of this limit is, again, a consequence of the continuity of the roots of an equation as a function of the parameters). With this definition, Bezout's theorem becomes evident.

21. In summary, Poncelet's idea is to "specialize" the parameters on which the objects under consideration depend in two ways, connecting, as it were, by "continuous variation" the given problem to another case of the same problem that has trivial solution, and to use a *principle of conservation of number*. This idea, considerably developed and diversified, is at the foundation of the research in *enumerative geometry* that, under Zeuthen and, above all, Schubert, continues the work of Chasles and of de Jonquières (IV, 13) in a much larger context. The Schubert calculus, for which he only gives justifications resting on "intuition," can, as Severi showed in substance in 1912, be considered as calculations in the *homology ring* of a nonsingular algebraic variety V. More precisely, since Schubert is interested only in numerical results, the notion of equivalence between algebraic cycles on V is not homological equivalence $A \underset{h}{\sim} B$ but what is called *numerical equivalence* $A \underset{n}{\sim} B$, which means that

$$(A.X) = (B.X)$$

for every algebraic cycle X on V whose dimension is complementary to the (equal) dimensions of A and B; it is clear that $A \underset{h}{\sim} B$ implies $A \underset{n}{\sim} B$.

22. Generally, Schubert considers a family of algebraic varieties that he can identify with points of an irreducible algebraic variety V in $\mathbf{P}_N(\mathbf{C})$ (cf. (VII, 31)); he considers "conditions" imposed on these varieties, which amounts to imposing conditions on the points identified with these varieties, and thus defines algebraic subvarieties W_j of V ($1 \leqslant j \leqslant r$). If the product cycle $W_1 . W_2 \ldots W_r$ is of dimension 0, its degree is, by definition, the number of varieties satisfying, simultaneously, all the required conditions. It is this number that Schubert tries to determine. To do that, he establishes, in various ways, relations between the classes of the W_j in the homology ring, which often allow him rapid approach to the objective.

23. For example, take for V, the grassmannian of the lines in the projective space $\mathbf{P}_4(\mathbf{C})$; let W_L, W_P, and W_H be the subvarieties of V composed of the lines that,

respectively, meet a line L, meet a plane P, and are contained in a hyperplane H. Then, Schubert establishes the relation:

$$W_P \,.\, W_P \underset{n}{\sim} W_L + W_H \tag{18}$$

by the following argument based on the "conservation of number": $W_P \sim W_{P_1}$, for any planes P, P_1; when P and P_1 are specialized to two planes that meet in a line L, then a line meeting P and P_1, meets L or else is contained in the hyperplane H generated by P and P_1, whence (18). From this it follows, in the homology ring:

$$W_P^6 \sim (W_L + W_H)^3 = W_L^3 + 3W_L^2 W_H + 3W_L W^2 + W_H^3. \tag{19}$$

But the cycle W_L^3 is of degree 1 (three arbitrary lines are met by only a single line), the cycle $W_L^2 W_H$ is of degree 1, the cycle $W_L W_H^2$ is of degree 0 (an arbitrary line and plane do not meet), and, finally, the cycle W_H^3 is of degree 1. Therefore, the relation (19) allows Schubert to state that in $\mathbf{P}_4(\mathbf{C})$ the number of lines meeting six planes in general position is 5.

24. In 1912, Severi made a comprehensive analysis of the validity of the "principles of conservation of number" used by his predecessors. The general form, under which this principle has been formulated since then, is that of a *correspondence* in a sense which generalizes that of (VI, 40), that is to say, a subvariety Γ of a product $V_1 \times V_2$ of two nonsingular projective varieties; it is assumed that the projection of Γ on V_1 is all of V_1 and that the dimensions of the irreducible components of Γ are at most equal to that of V_1. Then, for *every* simple point P of V_1, the cycle $(P \times V_2) \,.\, \Gamma$ is either of dimension $\geqslant 1$ or else of dimension 0 and its degree is *constant, provided that all the irreducible components of Γ have the same dimension* (equal to that of V_1).

The necessity for such a restriction is seen in the following example: take $V_1 = \mathbf{P}_4(\mathbf{C})$ and $V_2 = \mathbf{P}_3(\mathbf{C})$; a point with homogeneous coordinates $(t_0, t_1, t_2, t_3, t_4)$ in V_1 is identified with the polynomial:

$$P = t_0X^4 + t_1X^3 + t_2X^2 + t_3X + t_4$$

(defined up to nonzero factor), a point with homogeneous coordinates (a, b, c, d) in V_2 with the matrix

$$s = \begin{pmatrix} a & b \\ c & d \end{pmatrix}$$

(defined up to constant factor $\neq 0$), and the correspondence Γ is composed of the pairs (s, P) such that $s \,.\, P = P$ (up to nonzero factor). It is easily seen that if P is a polynomial with 4 distinct roots, then:

$$((P \times V_2) \,.\, \Gamma) = 4,$$

unless the cross ratio of these 4 roots (in a certain order) is -1, in which case the number is 8, or if this cross ratio is a cube root of unity, in which case the number is 12; this happens because Γ has several irreducible components of different dimensions.

25. Thus, the work of Severi and of Lefschetz emphasizes the essentially topological nature of the foundation of classical algebraic geometry; to be able to develop algebraic geometry over an arbitrary field in the same manner, it was necessary to invent purely algebraic tools that could be substituted for the topological notions of "continuous variation" and of "intersection number of cycles" (in the sense of algebraic topology). It is to van der Waerden that accrues the merit of having placed, beginning in 1926, the essential landmarks along this way.

To replace the idea of continuity, he first revives the process that had given birth to complex projective geometry, the *extension* of an algebraic variety in a space defined by a larger field, the idea whose very success made it vanish from sight, because of the richness of the structures supported by the field of complex numbers, and the techniques of analysis to which they gave access. In general, if K is an extension of a field k, and if an affine variety is defined as the set V of points of k^n satisfying a system of polynomial equations $P_\alpha(x_1, \ldots, x_n) = 0$ in the coordinates of this point (the P_α having their coefficients in k), then the *points of* K^n whose coordinates satisfy the same equations can also be considered, thus defining, starting from V, a variety V_K of K^n by "extension of scalars" from k to K. There is an analogous definition for projective varieties. That being so, van der Waerden's idea is to substitute the concept of *specialization* for that of continuity: given two extensions K, K′ of k, a *specialization* over k of an arbitrary finite system $(x_1, \ldots, x_m)$ of a finite number $m \geqslant 2$ of distinct elements of K is a map taking each x_j to a $x'_j \in K'$ (elements that need not be distinct), which are not all zero and which must satisfy the condition that $P(x'_1, \ldots, x'_m) = 0$ for every homogeneous polynomial $P \in k[T_1, \ldots, T_m]$ such that $P(x_1, \ldots, x_m) = 0$.

26. A point with homogeneous coordinates $(x_0, \ldots, x_n)$ in $\mathbf{P}_n(K)$, for a suitable extension K of k, is said to be a *generic point* of the projective variety $V \subset \mathbf{P}_n(k)$ if, for *every* extension K′ of k and *every* point with homogeneous coordinates $(x'_0, \ldots, x'_n)$ in $V_{K'}$, $(x'_0, \ldots, x'_n)$ is a specialization of $(x_0, \ldots, x_n)$. It is clear, then, that every property of the generic point, which is expressed by algebraic equations with coefficients in k, is still valid for *every* point of $V_{K'}$, thus specialization replaces the classical "passage to the limit."

Now, if $V \subset \mathbf{P}_n(k)$ is an *irreducible* variety, it is very easy to define a generic point of V; suppose for simplicity that k is *algebraically closed*; take for K the *field of rational functions* on V (the restrictions to V of the rational functions on $\mathbf{P}_n(k)$ defined at at least one point of V). As V cannot be contained in every hyperplane $x_j = 0$ $(0 \leqslant j \leqslant n)$, it may be assumed that it is not contained in the hyperplane $x_0 = 0$. Then, the restriction ξ_j to V of the rational function $x \mapsto x_j/x_0$ is defined at at least one point x of V, and, therefore, belongs to K. The definitions imply immediately that the point $\xi = (1, \xi_1, \ldots, \xi_n)$ is a generic point of V. Moreover, the existence of a generic point η with coordinates in a field K′ *characterizes* irreducible varieties (VI, 2), because if a product PQ of two polynomials is zero at every point, then $P(\eta)\,Q(\eta) = 0$, since k is infinite, thus $P(\eta) = 0$ or $Q(\eta) = 0$, and the definition of generic point implies that $P(x) = 0$ in V or $Q(x) = 0$ in V.

Furthermore, for all generic points η of V, the field $k(\eta)$ obtained by adjoining to k the nonhomogeneous coordinates of η is k-isomorphic to the field K, and its transcendence degree over k is the dimension of V.

27. These definitions, in spite of their apparent tautological character, permit purely algebraic proofs (therefore, valid in "abstract" algebraic geometry) to be given for numerous elementary theorems: for example, in an affine space k^N or a projective space $\mathbf{P}_N(k)$, the dimension of any irreducible component of a variety defined by d equations is at least equal to $N - d$ (a result that also follows from the "principal ideal theorem" proved in 1928 by Krull). From this it follows that if V and W are two irreducible varieties in $\mathbf{P}_N(k)$ (or in k^N), such that

$$d = \dim(V) + \dim(W) - N \geqslant 0,$$

then all the irreducible components of $V \cap W$ have dimension $\geqslant d$; those of dimension d are called *proper*.

28. The objective of intersection theory in "abstract" algebraic geometry is to associate to each *proper* component C of an intersection $V \cap W$ of two irreducible varieties an integer $\geqslant 0$, its *intersection multiplicity* $i(C, V.W)$, in such a way that a "calculus of cycles" can be established as in algebraic topology: a *r-cycle* will be a linear combination:

$$X = \sum_j a_j V_j$$

of irreducible varieties V_j of dimension r with positive or negative integer coefficients; if V and W are two irreducible varieties of dimensions r, s, respectively, and if all the irreducible components C_j $(1 \leqslant j \leqslant m)$ of $V \cap W$ are proper, then the cycle $V.W$ will be defined as equal to $\sum_j i(C_j, V.W) C_j$. Naturally, it is desired that this calculus have the formal properties analogous to those of the calculus of cycles in algebraic topology, particularly the associativity formula $X.(Y.Z) = (X.Y).Z$ when all the products are defined, and the projection formula $X.\mathrm{pr}_1(Y) = \mathrm{pr}_1(Y.(X \times k^m))$ when X is a cycle of k^n and Y a cycle of $k^n \times k^m$, provided that all the components of Y have the same dimension as their projection on k^n and behave in a "general" way at infinity. In the particular case where $\dim(X) + \dim(Y) = N$, where X, Y are cycles of $\mathbf{P}_N(k)$ and where all the components C of $X.Y$ are proper (therefore, reduced to a point), the intersection number $(X.Y)$ will be the sum $\sum_j i(C, X.Y)$ taken over all these components.

Finally, van der Waerden drew attention to the fact that intersection theory must give $i(C, V.W) = 1$ if a generic point of C (a component of $V \cap W$) is simple (VII, 41) on V and W and if the linear tangent varieties at this point intersect in the linear tangent variety to C ("criterion of multiplicity one"). In fact, A. Weil showed that if all these properties are satisfied for two different definitions of the function $i(C, V.W)$, these two definitions give the same number, provided that it is so for the special case where C is reduced to a point and V is a linear variety.

29. The first definition proposed for $i(C, V.W)$ (V, W irreducible subvarieties of $\mathbf{P}_N(k)$) was given by van der Waerden by a method, generalizing Poncelet's idea

(VII, 20), which transforms one of the varieties V, W using a degenerate linear transformation. Unfortunately, this method, essentially global in nature, only gave the definition of intersection multiplicity when not only C but *all* the other components of $V \cap W$ were proper. By two different methods, A. Weil and C. Chevalley showed how to avoid this drawback (for V and W irreducible subvarieties of k^N) by using the fact that C is the first projection of a component C′ of $\Delta \cap (V \times W)$ in $k^N \times k^N$, where Δ is the diagonal. Chevalley's definition (in the form given by Samuel a little later) clearly displays the *local* character of the intersection multiplicity: the local ring $\mathfrak{o}$ of C′ relative to $V \times W$ (VII, 39) and the ideal $\mathfrak{q}$ in $\mathfrak{o}$ generated by the differences $x_i - x_{i+N}$ of the coordinates in k^{2N} $(1 \leqslant i \leqslant N)$ are considered. Samuel shows that the length of the ring $\mathfrak{o}/\mathfrak{q}^m$ is finite and, for the integer m sufficiently large, is equal to a polynomial in m ("Hilbert-Samuel polynomial," cf. (VII, 17)) whose highest degree term is $e(\mathfrak{q})m^d/d!$, where d is the dimension of C′ and $e(\mathfrak{q})$ is an integer $\geqslant 0$; it is $e(\mathfrak{q})$ that is, by definition, the multiplicity $i(C, V . W)$.

30. For the case of plane curves the definition, given in the nineteenth century, of the intersection number $i(P, V . W)$ at a point P common to two irreducible curves V, W can be formulated in terms of the theory of local rings; if A is the local ring of k^2 at the point P, and $\mathfrak{p}_V$ and $\mathfrak{p}_W$ the prime ideals of A corresponding to V and W (VII, 39), $i(P, V . W)$ is the *length* of the A-module $A/(\mathfrak{p}_V + \mathfrak{p}_W)$. This definition can be carried over to the general case by replacing A by the local ring of C relative to k^N, and it is tempting to think that $i(C, V . W)$ would still be given by the preceding expression. It is exactly so when V and W are *complete intersections* of hypersurfaces (in number $N - \dim(V)$ and $N - \dim(W)$, respectively), but Gröbner has constructed examples where $i(C, V . W) \neq \text{length}(A/(\mathfrak{p}_V + \mathfrak{p}_W))$. In 1957, Serre showed how "to correct" this formula to obtain the value of the intersection number; it is necessary to use the "Tor" functors of homological algebra, which give the expression:

$$i(C, V . W) = \sum_{i=0}^{N} (-1)^i \text{length}(\text{Tor}_i^A(A/\mathfrak{p}_V, A/\mathfrak{p}_W)). \tag{20}$$

An interesting feature of this formula is that it can be extended to the theory of schemes.

31. To have the calculus of cycles render the same services as the homological methods in the classical case, it is still necessary to define the notion of *equivalence* of cycles that will substitute for homological equivalence. Since the beginning of the twentieth century, in the theory of algebraic surfaces over **C**, the Italian geometers used the notion of "(nonlinear) continuous system" of curves on a surface, meaning by that, a family of curves depending "algebraically" on a certain number of parameters. Generalizing ideas of Cayley (IV, 9) and of Bertini, Chow and van der Waerden showed, in 1937, how to parametrize the set of irreducible algebraic subvarieties V of a projective space $\mathbf{P}_N(k)$. We assume that V has dimension r; since a hyperplane of $\mathbf{P}_N(k)$ is identified with a point of another

projective space $\mathbf{P}_N(k)$ (the "dual" of the initial space $\mathbf{P}_N(k)$), sequences of $r+1$ hyperplanes are identified, therefore, with points $(\mathbf{u}^{(0)}, \ldots, \mathbf{u}^{(r)})$ of $(\mathbf{P}_N(k))^{r+1}$. The set S(V) of the sequences formed of $r+1$ hyperplanes *whose intersection meets* V can be shown to be a *hypersurface* in $(\mathbf{P}_n(k))^{r+1}$; in other words, it is defined by a single equation $F_V(\mathbf{u}^{(0)}, \ldots, \mathbf{u}^{(r)}) = 0$ in the coordinates of the hyperplanes $\mathbf{u}^{(0)}, \ldots, \mathbf{u}^{(r)}$, and F_V is an irreducible homogeneous polynomial in the $N+1$ homogeneous coordinates of each of the $\mathbf{u}^{(j)}$. F_V is said to be the *associated form* to V (defined up to a constant factor), and it can be proved that if V is of degree d, then F_V is homogeneous of degree d with respect to each of the systems of coordinates of the $\mathbf{u}^{(j)}$. The coefficients of F_V are called the *Chow coordinates* of V and, thus, associated to V is a well-determined point (Chow point) of a projective space $\mathbf{P}_\nu(k)$, where the dimension ν depends on r and d. The essential property of the Chow coordinates is that if U is a subvariety of $\mathbf{P}_N(k)$, the map that takes each algebraic subvariety V of U of dimension r and of degree d to its Chow point is *injective*, and its image is an *algebraic subvariety* $C_{r,d}(V)$ *of* $\mathbf{P}_\nu(k)$. The definition of the associated form and of the Chow coordinates is extended immediately to a *positive* r-cycle $X = \sum_j a_j V_j$ (the a_j positive integers) by taking the product $\prod_j (F_{V_j})^{a_j}$ for the associated form, the Chow coordinates of X being the coefficients of this form. Finally, for an arbitrary r-cycle $X = X' - X''$, where X' and X'' are positive, the Chow point of X is the pair of Chow points of X' and X''. Further, since the Chow points of the positive r-cycles $X = \sum_j a_j V_j$ of fixed *degree* $d = \sum_j a_j d_j$ (d_j, the degree of V_j), linear combinations of subvarieties of U, form an *algebraic variety*, so also do the Chow points of r-cycles $X' - X''$ where the degrees of the positive cycles X' and X'' are fixed.

32. Then, a *specialization* of a cycle can be defined (Matsusaka, Samuel) by considering a specialization of its Chow point (VII, 25) and the cycle whose Chow point is equal to this specialization. An algebraic family of r-cycles is composed of the cycles whose Chow points describe an algebraic variety, and if this variety is irreducible, the family is said to be irreducible. Two cycles X, Y are called *algebraically equivalent* if there exists a finite number of cycles $X_0 = X$, $X_1, \ldots, X_r = Y$ such that X_j and X_{j+1} belong to the same irreducible family for $1 \leqslant j \leqslant r-1$. If X, Y are algebraically equivalent to X', Y', respectively, and if the cycles $X.Y$ and $X'.Y'$ are defined, then they are algebraically equivalent (and, in particular, $(X.Y) = (X'.Y')$ if the dimensions of X and Y are complementary). This allows an equivalence class of cycles still to be defined even when the components of the intersection $X \cap Y$ are not all proper: it suffices to take the equivalence class of $X'.Y'$ where X' (resp. Y') is equivalent to X (resp. Y) and the intersection $X' \cap Y'$ has only proper components (which can be proved to be always possible).

4. Fields of definition and notions of algebraic variety

33. For a variety defined by algebraic equations with coefficients in a field k, the introduction of points whose coordinates are no longer in k but are in an extension

of k, such as the generic point (and that, even if k is algebraically closed), raises the question of the relations between the "geometric" notion of variety as a set of points and its "algebraic" aspect represented by its equations of definition.

In principle, in "abstract" algebraic geometry, algebraic equations with coefficients in any field whatsoever can be taken; but the experience of real geometry, where it is possible that *no* point of $\mathbf{R}^n$ satisfies a polynomial equation of degree > 0, shows that it would not be possible to have workable geometric statements without being over an algebraically closed field. Due to Hilbert's theorem of zeros (VII, 16) that is valid for any algebraically closed field whatsoever, every ideal of $k[T_1, \ldots, T_n]$ distinct from the unit ideal (resp. every graded ideal of $k[T_0, T_1 \ldots, T_n]$ not containing the "irrelevant" ideal generated by the T_j) has zeros in k^n (resp. $\mathbf{P}_N(k)$) *when* k is algebraically closed. Thus, all authors concerned with "abstract" algebraic geometry always consider (explicitly or not) that, for a system of algebraic equations (resp. homogeneous algebraic equations) over a field k, each point, whose coordinates (resp. homogeneous coordinates) are in the *algebraic closure* $\bar{k}$ of k and satisfy these equations, is a point of the "variety" defined by the equations under consideration.

34. Nevertheless, the fact that the equations of a variety have coefficients in a field *smaller* than the algebraically closed field on which the points of the variety are situated is a phenomenon that attracts more and more attention from 1940 on. Even at the height of the uncontested domination of complex projective geometry, a few works had explored the consequences that could be derived from this fact. It is the complex varieties whose equations have *real* coefficients that had been studied the most, especially questions concerning the topological properties of the sets of *real* points (i.e., having real coordinates) of the variety: in particular, questions of the maximal number of *circuits* of a real algebraic curve (Harnack), or the maximal number of circuits forming a sequence in which each is in the interior of the preceding (Hilbert); for example, a sextic can have at most 11 circuits but only two can be such that one of them is in the interior of the other.

Not only are the subfields containing the coefficients of the equations of a variety of interest but also the subfields containing the coefficients of the polynomials or rational functions entering in a birational transformation. It is thus that M. Noether remarked, as early as 1870, that if, on the coefficients of a birational transformation acting on a curve C, the condition of "not introducing irrationality," that is to say, of being in the field generated by the coefficients of the equations of C, is imposed, then a rational curve C cannot always be transformed into a line; if C is of even order, it can, in general, only be transformed into a conic.

35. Finally, after 1890, the first work on *diophantine geometry* begins. Here, (complex) algebraic varieties defined by equations with *rational* coefficients (resp. belonging to an algebraic number field) are considered and the properties of the points of the variety with rational or integer coordinates (resp. belonging to the given number field or its ring of integers) are studied. If the first work in this

direction, that of Hilbert and Hurwitz in 1890, is limited to genus 0 and remains rather superficial, Poincaré's memoir of 1901, devoted to the curves of genus 1, goes much further by systematically using the representation of the curve by elliptic functions (VII, 57). Using analogous methods, Mordell in 1922 for genus 1, and Weil in 1928 for curves of arbitrary genus, will prove that the group of rational points on the curve (considered as embedded in its jacobian (VII, 57)), is generated by a *finite* number of points, a result that Siegel, in 1929, will combine with his theorem of diophantine approximation to prove that a curve of genus $\geqslant 1$ in $\mathbf{C}^2$, defined by an equation with coefficients in the ring A of integers of an algebraic number field, has only a *finite* number of points with coordinates in A.

36. Thus, around 1940, it is always specified that a variety (affine or projective) is *defined over a field k* (or is a *k-variety*), that is to say, is defined by a system of equations with coefficients in k; similarly, the field of definition of a *correspondence* and, in particular, of a rational *function* has meaning by considering the subvariety of a product of projective spaces that corresponds to it. Finally, a *cycle* is defined over a field k if the varieties of which it is a linear combination are so defined.

In 1937, Chow and van der Waerden do not treat an algebraic k-variety V as a *set* of points but define only the notion of "point of the variety over a field K, an extension of k"; in set-theoretic language, this amounts, in substance, to *associating* to each of these fields K, the set V_K of points of K^n (or $\mathbf{P}_n(K)$) whose coordinates satisfy the equations of V (a point of view that will be made explicit in the theory of schemes (VIII, 42)), and the inclusion relation $V \subset W$ signifies that $V_K \subset W_K$ for all K.

In his 1946 work, *Foundations of Algebraic Geometry*, A. Weil adopts another point of view in order to be able to consider an algebraic variety as a *set* of points (although, in fact, for secondary reasons, he distinguishes a variety from the set of its points), and, at the same time, to have enough available "generic points" in this set. For this, he remarks that Hilbert's finite basis theorem implies that an affine or projective variety always has a field of definition generated by a *finite* number of elements over the prime field; thus, he restricts to fields of definition k that are extensions of finite type of the prime field k_0, and are contained in an algebraically closed extension Ω of k_0 of infinite transcendence degree over k, fixed once and for all, and called the *universal domain*. The points of a k-variety are only those whose coordinates are in Ω, and thus form a set that can be identified with the variety.

As for Zariski, he is, in his first work around 1940, closer to van der Waerden's position, and then he rallies to A. Weil's point of view, not without criticizing the abundance of "generic points," which he judges a little artificial (although in the classical case, the field **C** has exactly the properties of a "universal domain").

37. Whatever point of view is adopted, it is clear that every extension of a field k is also a field of definition of a k-variety. As the giving of a field of definition is an element of supplementary structure for a variety, one is led to the introduction of notions that make it intervene and to study the manner in which these notions

depend on the field of definition chosen. The simplest example refers to the notion of *irreducibility*. If V is an affine k-variety in K^n (where K is an algebraically closed extension of k but not necessarily a universal domain), the set of polynomials of $k[T_1, \ldots, T_n]$ that are zero on V is an ideal $\mathfrak{J}_k(V)$ and, following the model of the Kronecker-Lasker definition (VI, 2), to say that V is *irreducible over k* means that $\mathfrak{J}_k(V)$ is prime; it is easily shown to be equivalent to say that V is not the union of two distinct k-varieties of V. The *dimension* of the k-variety V is the transcendence degree *over k* of the field of fractions $R_k(V)$ of the integral domain

$$k[T_1, \ldots, T_n]/\mathfrak{J}_k(V).$$

Since the elements of this domain are identified with the restrictions to V of the polynomial maps

$$(t_1, \ldots, t_0) \mapsto P(t_1, \ldots, t_n),$$

where $P \in k[T_1, \ldots, T_n]$, $R_k(V)$ is called, once more, the *field of rational functions* on the k-variety V.

Now, if k' is an extension of k contained in K, corresponding to V, considered as a k'-variety, there is an ideal $\mathfrak{J}_{k'}(V)$ of $k'[T_1, \ldots, T_n]$; but even if $\mathfrak{J}_k(V)$ is prime, $\mathfrak{J}_{k'}(V)$ is not necessarily so, as the example of the variety defined over **R** and composed of the two points $\pm i$ of **C** shows. This example is characteristic, in the sense that if k' is taken equal to the algebraic closure $\bar{k}$ of k, a variety V irreducible over k, considered as $\bar{k}$-variety, has a finite number of irreducible components over $\bar{k}$, of the *same dimension* as V, that are permuted among themselves by the Galois group of $\bar{k}$ over k. V is said to be *absolutely irreducible* if, considered as a $\bar{k}$-variety, it remains irreducible; then for *every* extension k' of k, V is irreducible as a k'-variety.

38. The concept of a k-variety as a set of points allows A. Weil to free himself, for the first time, from projective geometry by defining what he calls "abstract varieties" or "Varieties." For the applications he had in mind (see (VII, 60)), he needed to construct algebraic varieties by a procedure analogous to the "gluing" of manifolds in algebraic topology or in differential geometry, which was well known in these theories from the first quarter of the twentieth century. He showed that it was possible to copy these procedures in algebraic geometry by taking for the "transition functions" (VII, 1) the birational and biregular maps of complements of subvarieties in affine k-varieties (which, later, will be called "Zariski open sets"). An important new notion introduced by A. Weil is that of *complete* (abstract) variety, which corresponds in algebraic geometry to the notion of compact manifold in topology. In classical projective geometry in $\mathbf{P}_N(\mathbf{C})$, all the algebraic varieties considered are complete, and A. Weil emphasized that it is this property that, in reality, allows the existence of *global* properties of projective varieties that affine varieties do not possess (affine varieties are never complete except when they are reduced to finite sets of points).

We note also that the theory of intersections and of cycles generalizes, by replacing k^N or $\mathbf{P}_N(k)$ by an arbitrary nonsingular abstract variety; it must be noted that in this theory the irreducible subvarieties considered are always assumed *absolutely irreducible*.

5. Local rings and their completions in algebraic geometry

39. The notion of *local ring* at a point P of an affine k-variety $V \subset \Omega^n$, irreducible over k, is introduced naturally with the notion of field of rational functions $R_k(V)$ (VII, 37): it is the subring of the functions *defined at the point* P, in other words, functions of the form u/v, where u and v are the restrictions to V of polynomial functions such that $v(P) \neq 0$. In this ring, denoted $\mathfrak{o}_k(P; V)$ or $\mathfrak{o}_k(P)$, the functions that are zero at the point P form the *unique* maximal ideal $\mathfrak{m}_k(P; V)$. The other prime ideals of $\mathfrak{o}_k(P; V)$ correspond one-to-one to the irreducible k-subvarieties of V that contain P. The polynomials of $k[T_1, \ldots, T_n]$ that are zero at P have, for common zeros in Ω^n, the points of an irreducible k-subvariety W of V of which P is a generic point (and which is sometimes called the *locus* of P); again, it is said that $\mathfrak{o}_k(P; V)$ is the local ring of V along W, and it can be said that it is the subring of $R_k(V)$ consisting of the functions defined at at least one point of W; it is also denoted by $\mathfrak{o}_k(W; V)$ and its maximal ideal by $\mathfrak{m}_k(W; V)$. The residue field $\mathfrak{o}_k(W; V)/\mathfrak{m}_k(W; V)$ is identified with the *field of rational functions* $R_k(W)$ on W. The *dimension* of the local ring $\mathfrak{o}_k(P; V)$ is, by definition, $\dim(V) - \dim(W)$. It can be shown that it is also the smallest number of elements necessary to generate an ideal of $\mathfrak{o}_k(P; V)$ primary for $\mathfrak{m}_k(P; V)$. Finally, it is the largest number of nonzero prime ideals of $\mathfrak{o}_k(P; V)$ in a strictly increasing chain $\mathfrak{p}_1 \subset \mathfrak{p}_2 \subset \ldots \subset \mathfrak{p}_r$ (which correspond to a strictly decreasing sequence of irreducible subvarieties of V containing W and distinct from V).

40. From 1927–1928 on, Grell and, especially, Krull had undertaken a study (essentially algebraic, without much contact with geometry) of the *general* notion of local ring, defined as having a *single maximal ideal*. A discrete valuation ring (VI, 10) over an *arbitrary* field is a local ring, and these rings (as well as their completions) had been at the foundation of Hensel's theory of *p-adic numbers* and of their applications to algebraic number theory. In 1938, Krull obtained a whole series of remarkable results on noetherian local rings, and, a little later, Zariski saw that these results could be put to use in algebraic geometry over an arbitrary field.

41. First of all, the last two definitions of "dimension" of a local ring of the type defined in (VII, 39) apply to *any* noetherian local ring whatsoever; it is easy to see that if $\mathfrak{o}$ is a noetherian local ring, $\mathfrak{m}$ its maximal ideal, and $K = \mathfrak{o}/\mathfrak{m}$ its residue field, the K-vector space $\mathfrak{m}/\mathfrak{m}^2$ has dimension at least equal to the dimension d of $\mathfrak{o}$. Krull studied in detail the local rings (now called *regular*) for which $d = \dim_K(\mathfrak{m}/\mathfrak{m}^2)$. It follows immediately that, in classical algebraic geometry, to say P is a *simple* point on a variety V signifies exactly that the local ring $\mathfrak{o}_{\mathbf{C}}(P; V)$ is regular, and Zariski proposed taking as definition of a simple point P on an arbitrary k-variety V the fact that $\mathfrak{o}_k(P; V)$ is a regular local ring. This definition has the advantage (even in the classical case) of being independent of the chosen embedding of V in an affine space, contrary to the usual jacobian criterion expressing that P is simple on a variety V, defined by the polynomial equations

$$F_\alpha(x_1, \ldots, x_n) = 0$$

in $\mathbf{C}^n$, by the fact that the jacobian of the F_α is of rank $n - \dim(V)$ at the point P. In fact, for an arbitrary field k, this criterion still implies that P is simple (in Zariski's sense), but the converse is true only if $R_k(V)$ is a *separable* extension of k. For example, if k is an imperfect field of characteristic $p > 0$, all the points of the curve with equation $x_1^p + x_2^p - a = 0$, where $a^{1/p} \notin k$, are simple in Zariski's sense, although the partial derivatives of $x_1^p + x_2^p - a$ are identically zero. Zariski showed how to modify the jacobian criterion by introducing other derivations of $R_k(V)$ that are zero in k^p but not in k; this allowed him to show that the singular points of V form an algebraic subvariety (in general reducible) of V of dimension $\leqslant \dim(V) - 1$.

42. The local ring at a point of a k-variety V depends on the field of definition k, and it is the same for the notion of simple point. For example, with the same notation as above, the point $(0, a^{1/p})$ on the plane curve $x_1^2 + x_2^p - a = 0$ (where $p > 2$) is simple when the curve is considered as k-variety but not if it is considered as $\bar{k}$-variety. For an absolutely irreducible variety V defined by the equations $F_\alpha = 0$, the usual jacobian criterion expresses that a point P remains simple when V is considered as $\bar{k}$-variety; then P is said to be *absolutely simple*.

43. Following the work of Dedekind in number theory, E. Noether and Krull, from 1927 on, had made a deep study of the notion of *integral* element over a commutative ring A (that is, an element satisfying a polynomial equation with coefficients in A whose highest degree term has coefficient 1) and of the notion of *integrally closed* integral domain (that is, a domain A whose field of fractions contains no element integral over A other than the elements of A). As early as 1939, Zariski discovered that the property, for a local ring $\mathfrak{o}_k(P; V)$ of a point P on a k-variety V, to be integrally closed (in which case it is said that P is *normal*) has remarkable consequences. A simple point is always normal but there are singular normal points, for example, the vertex of a cone of the second degree. Nevertheless, if an irreducible variety of dimension n is *normal* (that is, every point is normal), then the singular points form a variety of dimension *at most* $n - 2$; in particular, a normal curve is nonsingular. For a normal, irreducible projective variety V in $\mathbf{P}_N(k)$, the intersection of V and of a generic hyperplane of $\mathbf{P}_N(k)$ is also irreducible and normal (theorem of Seidenberg).

Zariski also showed that the notion of divisor (VI, 43) and its principal properties can be extended to *normal* varieties. But the most important result on normal varieties is what is called Zariski's "main theorem": if T is a birational transformation between a normal variety V and a variety V′, then there corresponds to a point P of V only one point of V′ *or* a subvariety of V′ *all* of whose irreducible components are of dimension $\geqslant 1$.

44. The interest taken by Zariski in normal varieties in the course of this early research stems, above all, from the fact that he found there the means to give a proof of the *resolution of singularities* of a k-surface over an arbitrary field of

characteristic 0. Indeed, he shows that for *every* irreducible k-variety V, a *normal* k-variety V′ can be defined, in an essentially unique way, as well as an everywhere defined birational transformation $T: V' \to V$ such that, if $T(P') = P$, then the local ring $\mathfrak{o}_k(P'; V')$ is the localization at P′ of the *integral closure* of the local ring $\mathfrak{o}_k(P; V)$. This process of *normalization* applied to an arbitrary irreducible algebraic plane *curve* C, already gives a nonsingular curve C′; a singular point of C will be, in general, the image under T of many simple points of C′. Note also that if k is algebraically closed, then for each singular point P of C, if $\mathfrak{o}'_k(P)$ is the integral closure of the local ring $\mathfrak{o}_k(P)$, the dimension of the quotient $\mathfrak{o}'_k(P)/\mathfrak{o}_k(P)$, considered as k-vector space, gives the "true" multiplicity at P without there being need to appeal to the notion of "infinitely near multiple points" (VI, 21). In fact, the genus of C is given by the formula analogous to M. Noether's (VI, 17, formula (21)):

$$g = \frac{1}{2}(n-1)(n-2) - \sum_{P \in C} \dim_k (\mathfrak{o}'_k(P)/\mathfrak{o}_k(P)) .$$

Applied to an algebraic *surface* V, normalization already gives a surface V′ having only *isolated* singular points, thus eliminating singular curves that were one of the stumbling blocks of the earlier attempts at an algebraic proof of the theorem (Walker's proof is transcendental in nature).

To remove the isolated singular points, Zariski takes up again the idea of applying a sequence of quadratic transformations (VI, 20), which he begins by simplifying, by showing that it can be arranged to "blow up" only a single point of the surface, and he substitutes the use of *general valuations* on a field K, introduced and studied in detail by Krull in 1931, for the consideration of "infinitely near multiple points." The only difference between the definition of a general valuation and of a discrete valuation is that the valuation takes its values in an *arbitrary* totally ordered group. Zariski made a penetrating study of the diverse valuations on $K = R_k(V)$ when V is a surface. The most interesting (and those that play the most important role in his proof) are obtained by "successive blow ups": a point $P_0 \in V$ is blown up; if V_1 is the corresponding surface and V'_1 its normalization, a point P_1 on the curve of V'_1 corresponding to P_0 is taken; then P_1 is blown up, and this process is repeated indefinitely. If $A_0, A_1, A_2, \dots$ are the local rings at the points $P_0, P_1, P_2, \dots$, they form an increasing sequence whose union, A, is the ring of a valuation over K with values in $\mathbf{R}$. It is through the study of this valuation that Zariski proves that if the process starts at a singular point P_0 and if the surfaces obtained from the singular points $P_1, P_2, \dots$ are blown up each time, then the process finishes by arriving at a curve on a V'_m, all of whose points are simple, for m large enough, from which he deduces, finally, the resolution of singularities.

It must be noted that if, after each blow up, the "normalization" of the surface obtained is omitted, it can happen, as B. Levi had noted, that the process never ends in simple points.

A little later, Zariski used analogous methods to resolve the singularities of a variety of dimension 3 over a field of characteristic 0; the proof is much more complicated, and, for this case, it is necessary to blow up, not only points but also curves on the variety. It is also by means of valuations that he was able to formulate, in purely algebraic terms, the problem of local uniformization of a variety (VI, 31) and to solve it when the base field is of characteristic 0.

45. It is precisely with regard to his work on local uniformization that Zariski introduced an idea that would have broad developments from 1950 on. It is necessary in this theory to consider the generalization of Dedekind-Weber's "abstract Riemann surface" (VI, 4), that is, the set S of *all* the valuations of a field of rational functions K that are zero on the base field. Using an idea introduced several years earlier (with regard to boolean rings) by M. H. Stone, Zariski defined a *topology* on S, for which S is a quasi-compact space, but not Hausdorff, in general; in the case of dimension 1, the closed sets of S are S itself and the finite subsets of S.

46. By an analogous procedure, a topology (now called the "Zariski topology") can be defined on every k-variety V: the closed sets are simply the k-subvarieties of V. It seems that Zariski had been led to introduce this notion by his desire to generalize to "abstract" algebraic geometry an important remark of Enriques, called, by Enriques, the "principle of degeneration," and that, for the usual topology of $\mathbf{P}_N(\mathbf{C})$ is almost trivial: if an irreducible algebraic variety $V \subset \mathbf{P}_N(\mathbf{C})$ depends continuously on a parameter and becomes reducible for certain values of this parameter, the "limit" variety for these values of the parameter remains *connected*, in any case. Now, the Zariski topology allows meaning to be given to this statement, the "passage to the limit" being replaced by a "specialization" (of cycles), of course; but the proof of this principle, far from being immediate, could be obtained by Zariski only by an extremely difficult method based on a new notion that he introduced for this purpose, that of "functions holomorphic along a subvariety."

This notion uses the idea of *completion* of a local ring, which is at the foundation of Hensel's theory of p-adic numbers and which had been generalized by Krull to all noetherian local rings: in such a ring $\mathfrak{o}$, assumed to be a domain, the powers $\mathfrak{m}^n$ of the maximal ideal $\mathfrak{m}$ form a fundamental system of neighborhoods of 0 for a metrizable topology on the ring, and the completed ring $\hat{\mathfrak{o}}$ of $\mathfrak{o}$ for this topology can be formed. For the local rings $\mathfrak{o}_k(P; V)$ of algebraic geometry (k being an extension of finite type of the prime field and V an irreducible k-variety), the completion has no nilpotent elements (theorem of Chevalley) but is not necessarily a domain; nevertheless, it is so if the point P is normal (theorem of Zariski); V is said to be *analytically irreducible* at P if the completion of $\mathfrak{o}_k(P; V)$ is a domain. That being so, if W is a k-subvariety of V and if V is analytically irreducible at each point of W, Zariski defines a ring $\mathfrak{o}^*_W$ of "functions holomorphic along W" that can be conceived as taking, at each point $P \in W$, a value in the completion of $\mathfrak{o}_k(P; V)$; the fundamental criterion that assists him in proving the "connectedness theorem" is that W is connected for the Zariski topology if and only if $\mathfrak{o}^*_W$ is a domain. These notions and their use in algebraic geometry were to be considerably clarified in the theory of schemes (see (VIII, 39)).

6. Zeta functions

47. A. Weil's work on the foundations of "abstract" algebraic geometry was largely motivated by problems in number theory dating back to 1921. In his thesis, E. Artin had observed that algebraic congruences in two variables modulo a prime number p, that is, of the form $F(x, y) \equiv 0 \pmod{p}$, where F is a polynomial with integer coefficients, can be interpreted as algebraic equations over the prime field $\mathbf{F}_p = \mathbf{Z}/p\mathbf{Z}$; similarly, "higher congruences," in the sense of Dedekind, are algebraic equations over a *finite* field $\mathbf{F}_q$ (with $q = p^d$). Moreover, the analogy, already exploited by Dedekind and Weber, between finite extensions of the field $\mathbf{C}(X)$ and algebraic number fields is much closer when $\mathbf{C}$ is replaced by $\mathbf{F}_q$ because the residue fields of the valuations of a finite extension of $\mathbf{F}_q(X)$ are *finite* fields (extensions of $\mathbf{F}_q$) exactly as for the valuations of a number field (whereas in classical algebraic geometry they are equal to $\mathbf{C}$). Restricting to quadratic extensions of a field $\mathbf{F}_q(X)$, Artin developed, on Dedekind's model, the theory of ideals in such a field, and introduced, in particular, the analogue of the Riemann-Dedekind zeta function (in the theory of algebraic number fields):

$$(21) \qquad \sum_{\mathfrak{a}} |N(\mathfrak{a})|^{-s} = \prod_{\mathfrak{p}} (1 - |N(\mathfrak{p})|^{-s})^{-1}$$

where $\mathfrak{a}$ ranges over the set of integral ideals of K and $\mathfrak{p}$ over the set of prime ideals, and N designates the norm. He was able to prove properties for this function very analogous to those of the usual zeta function: functional equation, position of the poles, absence of zeros in $\mathscr{R}s = 1$, and, in fact, he showed that it is a meromorphic function much simpler than in the case of number fields because it is a *rational* function of $u = q^{-s}$.

48. Artin had formulated his results as arithmetician without introducing any geometric interpretation; a little later, F. K. Schmidt observed that it was possible to generalize Artin's theorems to all the algebraic extensions of $\mathbf{F}_q(X)$ and to give an account of them that is much more natural and simple, by following the model of the Dedekind-Weber theory and introducing *divisors* in place of ideals. Thus, K is viewed as the field of rational functions on a nonsingular *algebraic curve* Γ defined over the field $\mathbf{F}_q$ that can be assumed embedded in a projective space $\mathbf{P}_N(\Omega)$, where Ω is a "universal domain" of characteristic p. For each integer $m \geqslant 1$, the points of Γ, whose coordinates are in the unique extension $\mathbf{F}_{q^m}$ of $\mathbf{F}_q$ of degree m, form a finite set of cardinal N_m; then, the definition of the zeta function of K (or of Γ) can be taken to be:

$$(22) \qquad \frac{d}{du}(\log Z(u)) = \sum_{m=1}^{\infty} N_m u^{m-1} \qquad \text{with } Z(0) = 1.$$

This function is rational in u and can be written:

$$(23) \qquad Z(u) = P_{2g}(u)/(1-u)(1-qu)$$

where P_{2g} is a polynomial of degree $2g$, g being the genus of Γ. In addition, F. K.

Schmidt discovered the remarkable fact that the functional equation:

(24) $$Z(1/qu) = q^{1-g}u^{2-2g}Z(u)$$

could be deduced directly from the *Riemann-Roch theorem* for Γ.

49. About the same time, the theory of congruences had led several number theorists to try to evaluate N_1 as a function of q; they had obtained inequalities of the form:

(25) $$|N_1 - (q+1)| \leqslant Cq^{\alpha},$$

where C is independent of q and $\frac{1}{2} < \alpha < 1$, and they had noted that $\alpha = \frac{1}{2}$ would be the best possible result. About 1932, Hasse became interested in this problem and noted that the relation $\alpha = \frac{1}{2}$ was a consequence of what Artin had already called the "Riemann hypothesis" for zeta functions, namely, the fact that all the zeros of the polynomial P_{2g} are found on the circle $|u| = q^{\frac{1}{2}}$. Indeed, this fact immediately implies the inequality:

(26) $$|N_1 - (q+1)| \leqslant 2g \,.\, q^{\frac{1}{2}}.$$

In 1934, Hasse succeeded in proving this hypothesis for $g = 1$, by adapting ideas, drawn from the theory of complex multiplication of elliptic functions, to the case of finite fields; moreover, he and Deuring observed that, to extend the result to genus $g \geqslant 2$, it would be necessary to use the theory of correspondences between curves (VI, 40).

50. Actually, it is due to this theory that A. Weil succeeded, in 1940, in proving the "Riemann hypothesis" for the zeta functions attached to a curve of any genus g over a finite field $\mathbf{F}_q$. It is necessary to begin by extending the notion of "effective" correspondence, by defining, more generally, a correspondence between two nonsingular curves C_1, C_2 as a *divisor* Γ on the surface $C_1 \times C_2$ and no longer only as a curve on this surface (VII, 24). Therefore, the *additive group* of the correspondences between C_1 and C_2 has meaning; an *equivalence relation* is introduced on it, where a divisor is equivalent to zero not only if it is principal but if it is a linear combination with integer coefficients of a principal divisor and of *degenerate* effective correspondences (VI, 40). Thus, there is a quotient group of the classes of correspondences. If, in addition, $C_1 = C_2 = C$, this quotient group $\mathfrak{A}(C)$ is equipped with the structure of a noncommutative *ring*, where the multiplication is induced by the composition of correspondences, which is defined (for effective correspondences) in the same way as the composition of set-theoretic relations in $C \times C$ and which is extended to the divisors on $C \times C$ by linearity; the ring $\mathfrak{A}(C)$ has, as identity element, the class of the diagonal Δ of $C \times C$.

In the classical case, Hurwitz's formula (formula (28) of (VI, 41)) shows that the number:

(27) $$S(\Gamma) = m_1(\Gamma) + m_2(\Gamma) - (\Gamma \,.\, \Delta),$$

where $m_1(\Gamma) = (\Gamma \,.\, (P \times C))$ and $m_2(\Gamma) = (\Gamma \,.\, (C \times P))$ (P a generic point of C), is the *trace* of an endomorphism of $H_1(C, \mathbf{Z})$. It is easily seen that this number depends only on the *class* of Γ in $\mathfrak{A}(C)$; the usual properties of traces show that for two elements ξ, η of $\mathfrak{A}(C)$:

(28) $$S(\eta . \xi) = S(\xi . \eta),$$

and it follows from a theorem of Castelnuovo that if ξ is the class of Γ in $\mathfrak{A}(C)$ and ξ' that of the inverse correspondence Γ^{-1}, then $S(\xi . \xi') \geqslant 0$, with equality only if $\xi = 0$.

51. To apply these properties to the "Riemann hypothesis," A. Weil had to begin by generalizing them to curves C over an arbitrary field. He could define $S(\Gamma)$ by the formula (27) due to his "calculus of cycles" (VII, 28) but, at first, he does not try to identify $S(\Gamma)$ with a trace, and gives direct proofs of (28) and of Castelnuovo's inequality. His fundamental idea is, then, that the number N_m, which is to be evaluated, is none other than the intersection number $(F^m . \Delta)$, where F is the *Frobenius correspondence* that, to each point of C, associates its image under the automorphism of C corresponding to the automorphism $t \mapsto t^q$ of the algebraic closure of $\mathbf{F}_q$. Therefore, by definition, $S(F^m) = 1 + q^m - N_m$. Then, Weil writes the inequality $S(\xi . \xi') \geqslant 0$ taking for ξ the class of $a . \Delta + b . F^m$, a and b being arbitrary integers, and immediately obtains the relation:

(29) $$|N_m - q^m - 1| \leqslant 2g . q^{m/2}$$

that generalizes (26) and proves the "Riemann hypothesis."

Subsequently, Weil showed, in addition, that $S(\Gamma)$ can actually be written as the trace of an endomorphism and that P_{2g} is the characteristic polynomial of this endomorphism (VII, 61); the conjectures he set forth at this time have been one of the most active catalysts in current research (IX, 123). On the other hand, the "Riemann hypothesis" on $Z(u)$ allows proofs, in analytic number theory, of "the best possible" upper bounds, inaccessible until now by other ways, for example, the upper bound of a "Kloostermann sum":

(30) $$\left| \sum_{x=1}^{p-1} \exp\left(\frac{2\pi i}{p}(x + x^{-1})\right) \right| \leqslant 2p^{1/2}.$$

7. Equivalence of divisors and abelian varieties

52. Along with linear equivalence of divisors on a surface, corresponding to linear systems of curves on the surface, the Italian geometers had studied intensively the *algebraic* equivalence of divisors (VII, 32), corresponding to what they called "(nonlinear) continuous systems" of curves. In general, if V is a normal, irreducible k-variety of dimension n, and if $\mathbf{G}$ designates the additive group of the divisors on V (in other words, the algebraic $(n-1)$-cycles), three important subgroups are distinguished in $\mathbf{G}$:

(31) $$\mathbf{G} \supset \mathbf{G}_n \supset \mathbf{G}_a \supset \mathbf{G}_l$$

where $\mathbf{G}_l$ (resp. $\mathbf{G}_a$, $\mathbf{G}_n$) is the subgroup of divisors linearly equivalent to 0 (resp. algebraically equivalent to 0, numerically equivalent to 0 (VII, 22)). When k is the field $\mathbf{C}$ of complex numbers, it is also necessary to introduce homological equivalence of divisors (VI, 36), but Lefschetz showed that, for divisors, this equivalence is the *same* as numerical equivalence. The central problem in the theory of divisors is the study of the group $\mathbf{G}/\mathbf{G}_l$, denoted also by Pic(V) and called the Picard group of V, and, more particularly, the groups $\mathbf{G}/\mathbf{G}_a$, $\mathbf{G}_n/\mathbf{G}_a$, and $\mathbf{G}_a/\mathbf{G}_l$.

53. The study of the group $\mathbf{G}/\mathbf{G}_a$ begins with Picard's memoir of 1901 on simple integrals of the third kind on a surface S (with ordinary singularities) in $\mathbf{P}_3(\mathbf{C})$. He proved that there exists an integer $\rho \geqslant 0$ such that $\rho + 1$ arbitrary curves on S are always the logarithmic curves of an integral of the third kind (that is to say, that when an arbitrary value y_0 is given to the (nonhomogeneous) coordinate y, for example, then the logarithmic singularities of the integral over the curve C_{y_0}, which is the intersection of S and of the plane $y = y_0$, are at the points of intersection of this plane and the $\rho + 1$ given curves); on the contrary, ρ curves can always be found for which there is no integral of the third kind having this property.

A little later, Severi established that the number ρ is the rank of the commutative group $\mathbf{G}/\mathbf{G}_a$; in other words, there exist ρ curves $C_1, \ldots, C_\rho$, algebraically independent on S, such that for any other curve C there is a relation $nC + n_1C_1 + \cdots + n_\rho C_\rho \underset{a}{\sim} 0$, where the n_j and n are rational integers not all zero and $\underset{a}{\sim}$ designates algebraic equivalence. Then, using Poincaré's "normal functions" (VI, 48), Lefschetz obtained in 1921 a remarkable criterion for a cycle of (real) dimension 2 on S to be homologous to a divisor: it is necessary and sufficient that all the double integrals of the first kind have their periods along this cycle equal to 0. First of all, this gave him (taking into account the identity between numerical and homological equivalence) the inequality $\rho \leqslant R_2$ (the second Betti number) (VII, 9), then, the fact that the classes of $H_2(S, \mathbf{Z})$ that are torsion elements come from divisors; in other words, $\mathbf{G}_n/\mathbf{G}_a$ is the (finite) torsion group in $H_2(S, \mathbf{Z})$ and $\mathbf{G}/\mathbf{G}_n$, a subgroup of $H_2(S, \mathbf{Z})$, is a free group on ρ generators.

We note also that it is from this study that Lefschetz deduced the first complete proof of a theorem stated by M. Noether: every algebraic curve on a "general" surface of degree $d \geqslant 4$ in $\mathbf{P}_3(\mathbf{C})$ is a complete intersection of the surface and of another surface in the space (IX, 103).

Next, Severi generalized the preceding results by proving that for every irreducible variety in a $\mathbf{P}_n(\mathbf{C})$, the group $\mathbf{G}_n/\mathbf{G}_a$ is finite, and the group $\mathbf{G}/\mathbf{G}_n$ is a free group of finite type (isomorphic to a $\mathbf{Z}^\rho$). The first of these results has been generalized by Matsusaka and the second by Néron, to varieties defined over an arbitrary field.

Finally, we note that on a surface S, the bilinear form $(D, D') \mapsto (D \,.\, D')$ (the intersection number of two divisors), defined on $\mathbf{G} \times \mathbf{G}$, gives, by passage to the quotient, a nondegenerate symmetric bilinear form on $(\mathbf{G}/\mathbf{G}_n) \times (\mathbf{G}/\mathbf{G}_n)$; Hodge proved that its signature is $(1, \rho - 1)$, which is equivalent to the statement that if $\deg(D) = 0$, then $(D.D) \leqslant 0$, the equality $(D.D) = 0$ being possible only if $D \underset{n}{\sim} 0$. This theorem also has been generalized to surfaces over an arbitrary field. As Mattuck-Tate and Grothendieck have shown, Castelnuovo's inequality (VII, 50), which is the essential point in the proof of the "Riemann hypothesis" for curves over a finite field, can be deduced from it.

54. The structure of the group $\mathbf{G}_a/\mathbf{G}_l$ has been elucidated by the theory of *abelian varieties*, the form taken, since the beginning of the twentieth century, by the theory of abelian functions and of theta functions, which dates back to Jacobi and Riemann. It is known that Jacobi (as well as Gauss before him, in unpublished papers) had represented elliptic functions by quotients of certain *integral* functions, the theta functions of a complex variable z, which, by the addition of a period, reproduce themselves multiplied by an exponential factor e^{az+b}. The problem of inversion of ultra-elliptic integrals, in the form given by Jacobi (V, 5), led Göpel and Rosenhain, likewise, to define theta functions of two complex variables, and Weierstrass to define certain theta functions of an arbitrary number of variables; but it is Riemann who must be considered the founder of the general theory.

Starting from a Riemann surface S of genus g and from a base $(C_j)_{1 \leqslant j \leqslant 2g}$ of its homology in dimension 1, to g integrals of the first kind W_h $(1 \leqslant h \leqslant g)$ corresponds a $g \times 2g$ matrix whose $2g$ columns $\mathbf{w}_j = (\omega_{hj})_{1 \leqslant h \leqslant g}$ are the periods of the W_h corresponding to C_j $(1 \leqslant j \leqslant 2g)$; replacing the base (W_h) of the space of integrals of the first kind by another, Riemann shows that the matrix (ω_{hj}) can be assumed to have the form (called "normal"):

$$\begin{pmatrix} 1 & 0 & \cdots & 0 & \tau_{11} & \tau_{12} & \cdots & \tau_{1g} \\ 0 & 1 & \cdots & 0 & \tau_{21} & \tau_{22} & \cdots & \tau_{2g} \\ \cdots & \cdots & \cdots & \cdots & \cdots & \cdots & \cdots & \cdots \\ 0 & 0 & \cdots & 1 & \tau_{g1} & \tau_{g2} & \cdots & \tau_{gg} \end{pmatrix}$$

In addition, he shows that if U, V are two integrals of the first kind, then $\int U\,dV = 0$, and if $U = X + iY$ (X, Y real), then $\int X\,dY > 0$, where the integrals are taken on the boundary of S, rendered simply connected. First of all, this gives him the condition:

(32) $\tau_{hk} = \tau_{kh}$, the matrix $T = (\tau_{hk})$ is therefore symmetric;

if, moreover, T is decomposed, $T = T' + iT''$, into real and imaginary parts, it follows that:

(33) the real symmetric matrix T'' is *positive nondegenerate*.

Then these conditions allow him to form, on Jacobi's model, a series $\Theta(\mathbf{z}) = \Theta(z_1, \ldots, z_g)$, an integral function of g complex variables, of the form:

$$\sum_{\mathbf{m}} \exp\left(Q(\mathbf{m}) + 2\pi i(\mathbf{m}|\mathbf{z})\right),$$

where the vector $\mathbf{m}$ ranges over $\mathbf{Z}^g$ and Q is a nondegenerate negative quadratic form, such that:

$$\begin{aligned} &(34) \qquad \Theta(\mathbf{z}+\mathbf{w}_j) = \Theta(\mathbf{z}) && \text{for } 1 \leqslant j \leqslant g \\ &\qquad\qquad \Theta(\mathbf{z}+\mathbf{w}_j) = \exp(\mathrm{L}_j(\mathbf{z})) \,.\, \Theta(\mathbf{z}) && \text{for } g+1 \leqslant j \leqslant 2g, \end{aligned}$$

where L_j is an affine linear function. Finally, Riemann shows how the "abelian functions," which operate the "inversion," in Jacobi's sense, of the system of equations:

$$(35) \qquad \mathrm{W}_h(x_1) + \cdots + \mathrm{W}_h(x_g) = z_h \qquad (1 \leqslant h \leqslant g)$$

(that is, the rational and symmetric functions of the $x_j \in \mathrm{S}$), can be expressed by quotients of products of integral functions of $\mathbf{z}$ deduced from the series Θ by translation.

55. Riemann's results on abelian functions were expanded in the last part of the nineteenth century by numerous mathematicians, particularly Weierstrass, Frobenius, Poincaré, and Picard. After Castelnuovo's and Picard's work, the results of this theory are expressed in the language of *abelian varieties*: if, in $\mathbf{C}^g$, a lattice Δ (isomorphic to $\mathbf{Z}^{2g}$) is considered, then a meromorphic function defined in $\mathbf{C}^g$ and admitting the vectors of Δ as periods is identified with a meromorphic function on the *complex torus* $\mathbf{C}^g/\Delta$, which is a holomorphic Kähler manifold and a commutative group. Then, it is proved that the following conditions are equivalent:

1° There exists a meromorphic function on $\mathbf{C}^g/\Delta$ that is invariant under only a finite number of translations of this group.

2° There exists a holomorphic submanifold Z of $\mathbf{C}^g/\Delta$ of (complex) dimension $g-1$ that is invariant under only a finite number of translations.

3° $\mathbf{C}^g/\Delta$ is isomorphic (as a holomorphic manifold) to a (nonsingular) closed holomorphic submanifold of a projective space $\mathbf{P}_\mathrm{N}(\mathbf{C})$ and, by a theorem of Chow, such a variety is necessarily an *algebraic* subvariety of $\mathbf{P}_\mathrm{N}(\mathbf{C})$.

4° There is a base of the space $\mathbf{C}^g$ such that the lattice Δ has a base (over $\mathbf{Z}$) of $2g$ vectors $\mathbf{w}_j$ $(1 \leqslant j \leqslant 2g)$ that are the columns of a matrix of the form:

$$\begin{pmatrix} e_1^{-1} & 0 & \cdots & 0 & \tau_{11} & \tau_{12} & \cdots & \tau_{1g} \\ 0 & e_2^{-1} & \cdots & 0 & \tau_{21} & \tau_{22} & \cdots & \tau_{2g} \\ \cdots & \cdots & \cdots & \cdots & \cdots & \cdots & \cdots & \cdots \\ 0 & 0 & \cdots & e_g^{-1} & \tau_{g1} & \tau_{g2} & \cdots & \tau_{gg} \end{pmatrix}$$

where the e_h $(1 \leqslant h \leqslant g)$ are integers > 0 and the matrix $T = (\tau_{hk})$ satisfies conditions (32) and (33).

In that case, $\mathbf{C}^g/\Delta$ is called an *abelian variety*. Then the structure of algebraic variety on this manifold (compatible with its structure of holomorphic manifold) is unique, the meromorphic functions being identified with the rational functions on the variety. Moreover, every holomorphic submanifold Z of the type described in

2° is the variety of zeros of a theta function on $\mathbf{C}^g$, that is, a function such that:

$$\Theta(\mathbf{z}+\mathbf{w})/\Theta(\mathbf{z}) = \exp(L_{\mathbf{w}}(\mathbf{z}))$$

for all $\mathbf{w} \in \Delta$, where $L_{\mathbf{w}}$ is an affine linear function.

56. Among the principal properties of abelian varieties, we cite Poincaré's theorem of complete reducibility: if A is an abelian variety, B an abelian subvariety of A (i.e. the image in $A = \mathbf{C}^g/\Delta$ of a $\mathbf{C}$-vector subspace $E \subset \mathbf{C}^g$ of dimension h such that $\Delta \cap E$ has rank $2h$), then there exists an abelian subvariety C in A of dimension $g-h$ such that $B \cap C$ is a finite group.

To a complex torus E/Δ, where $E = \mathbf{C}^g$, is associated a *dual* complex torus E'/Δ' of the same dimension, where E' is the dual complex vector space of E and Δ' is the lattice of E' composed of the vectors $\mathbf{z}'$ such that $\langle \mathbf{z}, \mathbf{z}' \rangle \in \mathbf{Z}$ for all $\mathbf{z} \in \Delta$ (in other words, the "Pontrjagin dual" of the compact group E/Δ). The dual torus of E'/Δ' is canonically identified with E/Δ; if follows that if E/Δ is an abelian variety then so is E'/Δ'. In general, E/Δ and E'/Δ' are not isomorphic but only *isogenic* (an *isogeny* $A \to B$ is a surjective homomorphism with *finite* kernel; if there exists such a homomorphism, then there exists also an isogeny $B \to A$, and A and B are said to be *isogenic*).

An *endomorphism* of a complex torus E/Δ is obtained, by passage to quotients, from a bijective $\mathbf{C}$-linear map of E onto itself that transforms Δ into itself. They form a ring $\mathscr{A}$, and the algebra $\mathscr{A} \otimes_{\mathbf{Z}} \mathbf{Q}$ over the field $\mathbf{Q}$ of rationals is designated by $\mathscr{A}_{(\mathbf{Q})}$. It follows that if E/Δ is an abelian variety, then the $\mathbf{Q}$-algebra $\mathscr{A}_{(\mathbf{Q})}$ is *semi-simple*, and that the center of each of its simple components is either a totally real algebraic number field or a quadratic extension of such a field.

57. Then Riemann's results can be interpreted by saying, first of all, that if Γ is a nonsingular curve of genus g in $\mathbf{P}_N(\mathbf{C})$, the complex torus $J = \mathbf{C}^g/\Delta$, where Δ is the lattice generated by the "period-vectors" $\mathbf{w}_j$ (VII, 54) is an abelian variety of dimension g, which is called the *jacobian* of Γ. It is worth remembering that the abelian variety J is of a special type (curves of genus g depend (up to isomorphism) on $3g-3$ complex parameters, whereas abelian varieties of dimension g depend (up to isomorphism) on $\frac{1}{2}g(g+1)$ complex parameters); moreover, the dual J' of J is *isomorphic* to J (and not only isogenic to J).

The divisors on Γ algebraically equivalent to 0 are, in this case, divisors of *degree* 0. Such a divisor can be written $\partial\gamma$, where γ is a 1-chain on the Riemann surface of Γ, and corresponding to it is the class $\phi(D)$ in $J = \mathbf{C}^g/\Delta$ of the vector $(\int_\gamma dW_1, \int_\gamma dW_2, \ldots, \int_\gamma dW_g)$. Abel's theorem shows that if D is principal, then $\phi(D)$ is the zero element of the group J, and the converse easily follows. Finally, the map ϕ is surjective and, as it is a homomorphism, the group $\mathbf{G}_a/\mathbf{G}_l$ has, therefore, the structure of an abelian variety.

If P_0 is a fixed point of Γ, the map $P \mapsto \phi(P - P_0)$ of Γ into its jacobian J is injective; it follows that it identifies Γ to a subvariety of dimension 1 of its jacobian. The case $g = 1$ is particularly interesting, for then Γ (an "elliptic curve") can be *identified* with its jacobian; in this case, the endomorphisms of J are called *complex multiplications* in Γ; the rational functions on Γ are identified with *elliptic functions* with the given lattice as lattice of periods and thus, a parametric representation of Γ by elliptic functions, an idea dating back to Clebsch, is obtained.

58. The preceding results admit generalizations in several directions, which take place around 1950. First of all, let V be a Kähler manifold of complex dimension n, and let p be an integer such that $0 \leqslant p \leqslant n-1$. Poincaré duality allows the canonical identification of the vector space of real homology $H_{2p+1}(V, \mathbf{R})$ with the cohomology space $H^{2n-2p-1}(V, \mathbf{R})$, which is itself identified with the space $\mathbf{H}_{\mathbf{R}}^{2n-2p-1}$ of the real harmonic $(2n-2p-1)$-forms; furthermore, Hodge theory permits on the latter space, the definition of the structure of a *complex* vector space for which it becomes canonically isomorphic to a subspace sum of certain $\mathbf{H}^{r,s}$ with $r+s=2n-2p-1$. Thus, a structure of *complex torus* $J_p(V)$ is canonically defined on $H_{2p+1}(V, \mathbf{R})/H_{2p+1}(V, \mathbf{Z})$. Moreover, Poincaré duality shows that the complex tori $J_p(V)$ and $J_{n-p-1}(V)$ are *dual* to one another.

These general definitions are due to A. Weil, who also proved that if V is a nonsingular compact *algebraic variety*, the tori $J_p(V)$ are *abelian varieties*. Moreover, the definition of the homomorphism ϕ of (VII, 57) can be generalized: if X is any algebraic cycle of complex dimension p on V, algebraically (therefore also homologically) equivalent to 0, then $X = \partial\gamma$, where γ is a $(2p+1)$-chain. It follows that, for every real harmonic $(2p+1)$-form ω, the integral $\int_\gamma \omega$ depends only on the class ξ in $H^{2p+1}(V, \mathbf{R})$ of the form ω, and, as the map $\omega \mapsto \int_\gamma \omega$ is linear, it can be considered as an element $x(\xi)$ of $H_{2p+1}(V, \mathbf{R})$. If ξ is replaced by $\xi + \zeta$, where ζ is a $(2p+1)$-cycle with integer coefficients, $x(\xi)$ is replaced by $x(\xi) + \zeta$, therefore the class $\phi_p(X)$ of $x(\xi)$ modulo $H_{2p+1}(V, \mathbf{Z})$ is well determined. Thus, a homomorphism of the group $\mathbf{G}_a^{(p)}$ of algebraic cycles of complex dimension p equivalent to 0, into the abelian variety $J_p(V)$, is defined.

59. It is not known if the properties of Riemann's map ϕ (VII, 57) generalize to ϕ_p, except in the two extreme cases $p=0$ and $p=n-1$. For $p=0$, the abelian variety $J_0(V)$ is denoted A(V) and is called the *Albanese variety* of V (in fact, it had been introduced by Severi in 1913). It is isomorphic to $\mathbf{H}^{n-1,n}/H^{2n-1}(V, \mathbf{Z})$, and is characterized by the following "universal property": if P_0 is a fixed point of V and if ψ is defined by $\psi(P) = \phi_0(P - P_0)$, for every point $P \in V$, then every rational map $f: V \to B$, where B is an abelian variety, factors uniquely into $f = \lambda \circ \psi + c$, where c is a constant element of B and $\lambda: A \to B$ a homomorphism.

For $p = n-1$, the abelian variety $J_{n-1}(V)$ is called the *Picard variety* of V. It is isomorphic to $\mathbf{H}^{0,1}/H^1(V, \mathbf{Z})$. Since $\mathbf{G}_a^{(n-1)}$ is the group $\mathbf{G}_a$ of divisors algebraically equivalent to 0, the homomorphism ϕ_{n-1} maps $\mathbf{G}_a$ into $J_{n-1}(V)$; it follows that it defines an isomorphism of groups $\mathbf{G}_a/\mathbf{G}_l \to J_{n-1}(V)$. The abelian varieties $J_0(V)$ and $J_{n-1}(V)$ are dual to one another and are of (complex) dimension $\frac{1}{2}R_1$.

60. The theory of abelian varieties over an *arbitrary* field k has been entirely constructed by a single mathematician, A. Weil. He defines an abelian variety as a

complete (abstract) variety A on which is defined an *algebraic group* structure (i.e. such that the product $(x,y) \mapsto xy$ and the inverse $x \mapsto x^{-1}$ are rational maps); such a group is necessarily *commutative* (the analogue of the classical theorem stating that a *connected, compact complex* Lie group is necessarily commutative). His method consists essentially of working with the divisors on A and of studying the effect a translation of the group A has on a divisor, from the point of view of linear equivalence, by using an earlier construction of the jacobian of a curve as an auxiliary tool. Thus, he succeeds in generalizing to an arbitrary base field k, Poincaré's theorem of complete reducibility, and to prove that in the additive group A, the map $x \mapsto n\,.\,x$ is a surjective homomorphism for all integers n, with kernel A_n, a *finite* subgroup of n^{2g} elements (where $g = \dim(A)$), provided that n is not divisible by the characteristic p of k (a result which, when $k = \mathbf{C}$, is trivial for every complex torus $\mathbf{C}^g/\Delta$).

61. This fundamental result then allows Weil to study the homomorphisms of abelian varieties by their effect on "l-adic modules" that, for him, take the place of homology groups in the classical case. For a prime number $l \neq p$, the finite subgroups A_{l^m} of A form an inverse system whose inverse limit $T_l(A)$ is a free module of rank $2g$ over the ring $\mathbf{Z}_l$ of l-adic integers (now called the *Tate module* of A over $\mathbf{Z}_l$). Consequently, if $\alpha : A \to B$ is a homomorphism of abelian varieties, then, for $l \neq p$, there is a homomorphism $T_l(\alpha) : T_l(A) \to T_l(B)$ of $\mathbf{Z}_l$-modules canonically corresponding to it. Thus, in particular, to every endomorphism α of A is associated an endomorphism $T_l(\alpha)$ of the free $\mathbf{Z}_l$-module $T_l(A)$, and the characteristic polynomial of $T_l(\alpha)$ (of degree $2g$) has rational *integer* coefficients that are independent of the prime number $l \neq p$ chosen.

It is in this way that A. Weil obtains the interpretation of $P_{2g}(u)$ in his theory of the zeta function (VII, 51): the Frobenius correspondence F on the curve C canonically defines an endomorphism J(F) of the jacobian J(C) of C, and therefore, for each $l \neq p$, an endomorphism $T_l(J(F))$ of $T_l(J(C))$; P_{2g} is the characteristic polynomial of this endomorphism, and S(F) its trace. The same method allows him also to prove Artin's conjectures on the "L functions" associated to a finite group G of automorphisms of the curve C (defined over $\mathbf{F}_q$, as are the automorphisms in G). For each character χ of G, the function L_χ is defined by:

$$\log L_\chi(u) = \sum_{n=1}^{\infty} \nu_n(\chi)\, u^n/n \tag{36}$$

with

$$\nu_n(\chi) = (\mathrm{Card}(G))^{-1} \sum_{s \in G} \chi(s^{-1}) \Lambda(sF^n), \tag{37}$$

where $\Lambda(sF^n)$ is the number of fixed points of the automorphism sF^n of C. Artin had conjectured that L_χ is a meromorphic function of u (in fact, for the analogous functions defined for algebraic number fields); here, it is easy to see that L_χ can have only a finite number of singular points, and therefore, must be a *rational* function of u; this is what Weil proved using his theory of "l-adic matrices."

62. After the appearance of A. Weil's memoir, various mathematicians completed his results, bringing the theory of abelian varieties over an arbitrary field to the same point as the classical theory. Matsusaka, then Chow and Néron and Samuel gave definitions of the Albanese and Picard varieties of an arbitrary algebraic variety V, and showed that they have properties analogous to those described above for the case $k = \mathbf{C}$ (in particular, the group $\mathbf{G}_a/\mathbf{G}_l$ can be identified with the Picard variety). The dual A′ of an abelian variety A is defined as the Picard variety of A, and Cartier and Nakai were able to prove that the dual $(A')'$ is canonically identified with A. Finally, Weil, who had defined abelian varieties, at first, as "abstract varieties," showed that they can, in fact, he embedded in a projective space $\mathbf{P}_N(k)$.

VIII — THE SEVENTH EPOCH

Sheaves and Schemes
(1950–)

Since 1945, the considerable advances in algebraic topology, differential topology, and the theory of analytic spaces, due to the introduction of the notions of *sheaf*, of *cohomology with coefficients in a sheaf*, and of *spectral sequence* (all three invented by J. Leray), have also completely renovated the concepts and the methods of algebraic geometry, "classical" as well as "abstract." The language of cohomology permits the unification and simplification of many definitions and results, and opens the way for the solution of numerous problems.

1. The Riemann-Roch problem for varieties of dimension $\geqslant 2$ and the cohomology of sheaves

1. We have seen (VII, 43) that the notion of divisor and of linear equivalence and, in the classical case of varieties over **C**, the notion of canonical divisor can be defined without difficulty on a nonsingular, irreducible algebraic variety. The Riemann-Roch problem for a variety V of dimension $\geqslant 2$ consists in giving a formula expressing the dimension $l(\mathrm{D})$ of the space $\mathrm{L}(\mathrm{D})$ attached to an arbitrary divisor D on V using invariants attached to V or to D and the number $l(\Delta - \mathrm{D})$. The Italian geometers had studied this problem in detail for algebraic surfaces but they reached only an *inequality* for $l(\mathrm{D})$:

$$l(\mathrm{D}) \geqslant (\mathrm{D}\,.\,\mathrm{D}) - \pi(\mathrm{D}) + p_a + 1 - l(\Delta - \mathrm{D}), \tag{1}$$

where $\pi(\mathrm{D})$ is the "virtual genus" of an arbitrary curve of the linear system associated to D (VI, 45), by assuming for simplicity that $l(\mathrm{D}) > 0$ and that the curves of this linear system are irreducible.

2. In the period 1930–1940, the study of differential geometry and, in particular, of E. Cartan's method of moving frames, led ultimately to the notion of *fiber space* and, more particularly, of *vector bundle* on a differential manifold M. A vector bundle is a differential manifold $\mathbf{E}$ together with a projection $p : \mathbf{E} \to \mathrm{M}$, which is surjective and of class C^∞, such that for all $x \in \mathrm{M}$, the "fiber" $p^{-1}(x)$ is a real (resp. complex) vector space of fixed dimension r (the *rank* of $\mathbf{E}$); moreover, "locally on M," $\mathbf{E}$ behaves like a product $\mathrm{M} \times \mathbf{R}^r$ (resp. $\mathrm{M} \times \mathbf{C}^r$): more precisely, each point

of M has an open neighborhood U for which there is a diffeomorphism ϕ mapping $p^{-1}(U)$ into $U \times \mathbf{R}^r$ (resp. $U \times \mathbf{C}^r$) such that ϕ *linearly* transforms each fiber $p^{-1}(x)$ into $\{x\} \times \mathbf{R}^r$ (resp. $\{x\} \times \mathbf{C}^r$) (it is said that ϕ trivializes $\mathbf{E}$ over U). Thus, the notion of vector bundle corresponds to the intuitive idea of a vector space attached to each point of M and "varying differentiably" with this point. A *section* of $\mathbf{E}$ over an open set U is a map s of U into $\mathbf{E}$, of class C^∞, such that $s(x) \in p^{-1}(x)$ for all $x \in U$; the vector space of sections of $\mathbf{E}$ over U is denoted by $\Gamma(U, \mathbf{E})$ and simply by $\Gamma(\mathbf{E})$ when $U = M$. If M is a *holomorphic* manifold, *holomorphic* vector bundles on the base M can similarly be defined by requiring that $\mathbf{E}$ be a holomorphic manifold, that p be a holomorphic map, with the fibers $p^{-1}(x)$ complex vector spaces and the diffeomorphisms ϕ (in the definition above) holomorphic. As important examples of vector bundles, the *tangent bundle* $T(M)$, where $p^{-1}(x)$ is the space of tangent vectors at the point x (such that the rank of $T(M)$ is $\dim(M)$) and the *bundle of the p-cotangent vectors*, whose sections above M are the exterior differential p-forms (VII, 2), must be cited.

3. On the other hand, the concept of *divisor* can be generalized to any holomorphic manifold M, whatsoever. Let (U_α) be an open covering of M; consider on each U_α, a meromorphic function h_α such that, in $U_\alpha \cap U_\beta$ h_α/h_β is *holomorphic and* $\neq 0$ at every point. Two such systems (h_α), (h'_λ) corresponding to two coverings (U_α), (U'_λ) (not necessarily distinct) are identified if h_α/h'_λ is holomorphic and $\neq 0$ in $U_\alpha \cap U'_\lambda$ for every pair of indices (α, λ). These classes of systems are called *divisors* on M. For every meromorphic function f on M, h_α can be taken equal to $f|U_\alpha$ for all α, and the divisor (f), thus defined, is again said to be *principal*. It is easy to see that if M is a nonsingular projective algebraic variety, this notion coincides with that introduced in (VI, 43). For example, if $M = \mathbf{P}_n(\mathbf{C})$ and if $D = \sum_k m_k S_k$ is a divisor in the sense of (VI, 43), the m_k being positive or negative integers and each S_k a hypersurface defined by an irreducible, homogeneous equation $F_k(x_0, x_1, \ldots, x_n) = 0$ of degree d_k, then a divisor, in the sense defined above, is made to correspond to it in the following way. $\mathbf{P}_n(\mathbf{C})$ is covered by $n+1$ open sets $U_j (0 \leqslant j \leqslant n)$, where U_j is the open complement of the hyperplane $x_j = 0$, and, in U_j, one takes the meromorphic function:

$$h_j : x_j \mapsto x_j^{-d} \prod_k (F_k(x_0, x_1, \ldots, x_n))^{m_k} \tag{2}$$

where $d = \sum_k m_k d_k$.

4. In 1949, A. Weil observed that a *rank 1 holomorphic vector bundle* (also called a *line bundle*) $\mathbf{B}(D)$ is naturally attached to a divisor D on a holomorphic manifold M: with the notation of (VIII, 3), the holomorphic varieties $U_\alpha \times \mathbf{C}$ are "'glued" by taking the function:

$$(x, z) \mapsto (x, (h_\beta(x)/h_\alpha(x))z) \tag{3}$$

which is holomorphic in $(U_\alpha \cap U_\beta) \times \mathbf{C}$ as "transition function" from $U_\alpha \times \mathbf{C}$ to $U_\beta \times \mathbf{C}$ (VII, 1). In addition, if s is a *holomorphic section* of $\mathbf{B}(D)$, the restrictions $s_\alpha = s|U_\alpha$ satisfy $s_\beta = (h_\beta/h_\alpha)s_\alpha$ in $U_\alpha \cap U_\beta$ and, consequently, there exists a *mero-*

morphic function f on M such that the restriction $f|U_\alpha$ is equal to s_α/h_α for all α. For an algebraic variety M, it is equivalent to say that $(f) + D \geqslant 0$ and, consequently ((VI, 8) and (VI, 43)), the vector space L(D) can be interpreted as *the space* $\Gamma(\mathbf{B}(D))$ *of the holomorphic sections* of the line bundle $\mathbf{B}(D)$. For example, if $M = \mathbf{P}_n(\mathbf{C})$ and if $D = H$, an arbitrary *hyperplane* of $\mathbf{P}_n(\mathbf{C})$, then the transition functions for $\mathbf{B}(H)$ are (for the covering (U_j) defined in (VIII, 3)):

$$(x, z) \to (x, (x_k/x_j)z) \tag{4}$$

in $U_j \cap U_k$, and $\Gamma(\mathbf{B}(H))$ is precisely the vector space of *all the linear forms*

$$(x_0, x_1, \ldots, x_n) \mapsto \lambda_0 x_0 + \cdots + \lambda_n x_n$$

on $\mathbf{C}^{n+1}$.

Conversely, it follows that every holomorphic line bundle, with base a projective algebraic variety M, can be obtained (up to isomorphism) as a bundle of the form $\mathbf{B}(D)$; moreover $\mathbf{B}(D)$ and $\mathbf{B}(D')$ are isomorphic bundles if and only if D and D′ are *linearly equivalent* (VI, 43).

5. Between 1940 and 1950, the study of the topology of differential manifolds led to attaching to each *complex* vector bundle $\mathbf{E}$ on a differential manifold M of (real) dimension n, for each even integer $2j \leqslant n$, a well-determined element $c_j(\mathbf{E})$ of the cohomology group $H^{2j}(M, \mathbf{Z})$, called the *Chern class* of index j of $\mathbf{E}$. There are several ways to define them: they are *characterized* by the following properties, where $c(\mathbf{E})$ is set equal to $\sum_{j=0}^{\infty} c_j(\mathbf{E})$ (a finite sum since $H^{2j}(M, \mathbf{R}) = 0$ for $2j > \dim(M)$; by convention, $c_0(\mathbf{E}) = 1$).

1° The direct sum

$$\mathbf{E}_1 \oplus \mathbf{E}_2 \oplus \cdots \oplus \mathbf{E}_m$$

of complex vector bundles on M is defined in the obvious way by taking as trivializing diffeomorphism over a small enough open set, the direct sum of the trivializing diffeomorphisms for the $\mathbf{E}_j$. Then:

$$c(\mathbf{E}_1 \oplus \mathbf{E}_2 \oplus \cdots \oplus \mathbf{E}_m) = c(\mathbf{E}_1)c(\mathbf{E}_2)\cdots c(\mathbf{E}_m), \tag{5}$$

where the product is taken in the cohomology ring $H^\bullet(M, \mathbf{Z})$.

2° For every map $f: M' \to M$ of class C^∞, the "inverse image" $f^*(\mathbf{E})$ of a vector bundle $\mathbf{E}$ on M is defined as the subvariety of $M' \times \mathbf{E}$ composed of the pairs (x', z) such that $f(x') = p(z)$; it is a vector bundle with base M′. One has:

$$c(f^*(\mathbf{E})) = f^*(c(\mathbf{E})) \tag{6}$$

where, on the right side, $f^*: H^\bullet(M, \mathbf{Z}) \to H^\bullet(M', \mathbf{Z})$ is the homomorphism deduced canonically from f.

3° $$c(\mathbf{B}(H)) = 1 + h_n$$

for a hyperplane H of $\mathbf{P}_n(\mathbf{C})$, where $h_n \in H^2(\mathbf{P}_n(\mathbf{C}), \mathbf{Z})$ is the cohomology class corresponding, by Poincaré duality, to the $(2n-2)$-cycle H.

If M is a holomorphic manifold of *complex* dimension n, the Chern classes of the tangent bundle T(M) are simply written c_j $(1 \leqslant j \leqslant n)$ and called the *Chern classes* of M; when M is considered as a $2n$-cycle, it follows that

(7) $$\langle c_n, M\rangle = \chi(M) = \sum_{j=0}^{2n} (-1)^j R_j$$

(Euler-Poincaré characteristic of M).

6. In 1951, using the interpretation of divisors as line bundles, as well as Hodge theory, Kodaira showed that there is a "Riemann-Roch formula" for *compact Kähler manifolds* of dimension 2, generalizing the results of the Italian geometers: in the right side of (1), he replaces p_a by $p_g - \frac{1}{2}R_1$, which makes sense for every compact holomorphic manifold of dimension 2, and he expresses the difference between the two sides of (1) (the "superabundance" of $|D|$, in the Italian terminology) by "deficiencies" of the linear series on the generic curve of the system $|D|$ (generalizing a formula of Castelnuovo) (see (VIII, 17)). The following year, he obtained an analogous formula for compact Kähler manifolds of dimension 3; in both cases he had observed the role of the Chern classes in his formulas.

As early as 1909, Severi, while defining the arithmetic genus of a nonsingular irreducible projective algebraic variety of arbitrary dimension n, had conjectured that its arithmetic genus was given by:

(8) $$p_a = g_n - g_{n-1} + g_{n-2} - \cdots + (-1)^{n-1} g_1,$$

where g_j designates the dimension of the complex vector space of holomorphic differential j-forms on V. Inspired by the proofs of Severi and Zariski, Kodaira was able to prove this conjecture in 1952 using Hodge theory and a study of a particular case of the Riemann-Roch problem, which allowed him to calculate $l(\Delta + D)$ for certain divisors D.

7. Meanwhile, H. Cartan and Serre had discovered that the notion of *sheaf* introduced in 1945 by J. Leray permitted the expression of the fundamental results of the theory of holomorphic manifolds in a remarkably simple and suggestive form. Holomorphic functions, each defined on an open set (depending on the function) of a holomorphic manifold M, satisfy the sheaf axioms in the form given by H. Cartan: if $\mathcal{O}(U)$ designates the set of holomorphic functions defined in the open set U of M then, for every open covering (V_α) of U: (1) a function $f \in \mathcal{O}(U)$ is entirely determined by its restrictions $f|V_\alpha \in \mathcal{O}(V_\alpha)$, and conversely (2) if, for each α, a function $f_\alpha \in \mathcal{O}(V_\alpha)$ is given, such that for all pairs (α, β), f_α and f_β have the same restriction to $U_\alpha \cap U_\beta$, then there exists a function $f \in \mathcal{O}(U)$ such that $f|V_\alpha = f_\alpha$ for all α. The sheaf thus defined is called the *structure sheaf* of M and is denoted $\mathcal{O}_M$; $H^0(U, \mathcal{O}_M)$ is written in place of $\mathcal{O}(U)$. More generally, for every complex vector bundle $\mathbf{E}$ on M, a sheaf $\mathcal{O}(\mathbf{E})$ on M is defined by replacing, in the preceding definition, $\mathcal{O}(U)$ by the set $\Gamma(U, \mathbf{E})$, also written $H^0(U, \mathcal{O}(\mathbf{E}))$, of *holomorphic sections* of $\mathbf{E}$ over U. In particular, the sheaf corresponding to the vector bundle of the p-cotangent vectors of M is written Ω_M^p so that $H^0(U, \Omega_M^p)$ is the set of holomorphic p-forms on U. For a divisor D on M, $\mathcal{O}_M(D)$ is written in place of $\mathcal{O}(\mathbf{B}(D))$.

8. But there are types of sheaves other than those coming from vector bundles, and the great interest of sheaves lies in this diversity and the flexibility resulting from it. Indeed, we observe that in the axioms of (VIII, 7) the only property that intervenes is the fact that an element of $\mathcal{O}(\mathrm{U})$ has a well determined *restriction* to an open set $\mathrm{V} \subset \mathrm{U}$, in other words, that there is a linear map $\mathcal{O}(\mathrm{U}) \to \mathcal{O}(\mathrm{V})$ such that if W is an open set and $\mathrm{W} \subset \mathrm{V} \subset \mathrm{U}$, then the map $\mathcal{O}(\mathrm{U}) \to \mathcal{O}(\mathrm{W})$ is the composite of $\mathcal{O}(\mathrm{V}) \to \mathcal{O}(\mathrm{W})$ and of $\mathcal{O}(\mathrm{U}) \to \mathcal{O}(\mathrm{V})$. Thus, in the axioms, $\mathcal{O}(\mathrm{U})$ can be replaced by a group $\mathscr{F}(\mathrm{U})$ (which, generally, will be a $\mathcal{O}(\mathrm{U})$-module), with these groups having "restrictions" (homomorphisms $\mathscr{F}(\mathrm{U}) \to \mathscr{F}(\mathrm{V})$) analogous to those of $\mathcal{O}(\mathrm{U})$. In this way, the definition of a *sheaf of groups* $\mathscr{F}$ is obtained; $\mathrm{H}^0(\mathrm{U},\mathscr{F})$ is written in place of $\mathscr{F}(\mathrm{U})$, and the elements of $\mathrm{H}^0(\mathrm{U},\mathscr{F})$ are called the *sections* of $\mathscr{F}$ over U. Then, to each point $x \in \mathrm{M}$ is associated the *fiber* $\mathscr{F}_x$ of the sheaf at this point, which can be defined in the following way: the elements of $\mathscr{F}_x$ are the equivalence classes for the relation "$s \in \mathscr{F}(\mathrm{U})$ and $s' \in \mathscr{F}(\mathrm{V})$ have the same restriction to a sufficiently small $\mathrm{W} \subset \mathrm{U} \cap \mathrm{V}$," for U, V, W open neighborhoods of x ("germs of sections at x").

To all the usual operations on groups or modules correspond operations on sheaves. For example, a sheaf $\mathscr{N}$ is contained in a sheaf $\mathscr{G}$ (or is a *subsheaf* of $\mathscr{G}$) if $\mathscr{N}(\mathrm{U})$ is a subgroup of $\mathscr{G}(\mathrm{U})$ for all open sets U. Then (if $\mathscr{N}$ is normal in $\mathscr{G}$) a quotient sheaf $\mathscr{G}/\mathscr{N}$ is defined; it has the property that $(\mathscr{G}/\mathscr{N})_x = \mathscr{G}_x/\mathscr{N}_x$. Each fiber $(\mathcal{O}_{\mathrm{M}})_x$, denoted $\mathcal{O}_x$, is a *local ring* (VII, 40). If the sheaves $\mathscr{F}$ and $\mathscr{G}$ are $\mathcal{O}_{\mathrm{M}}$-Modules (which means that $\mathscr{F}(\mathrm{U})$ and $\mathscr{G}(\mathrm{U})$ are $\mathcal{O}(\mathrm{U})$-modules for every open set U), the sheaf $\mathscr{F} \otimes_{\mathcal{O}_{\mathrm{M}}} \mathscr{G}$ (also written $\mathscr{F} \otimes \mathscr{G}$), whose fiber at each point $x \in \mathrm{M}$ is $\mathscr{F}_x \otimes_{\mathcal{O}_x} \mathscr{G}_x$, is defined. It follows, in particular, that for two divisors D, D′:

$$(9) \qquad \mathcal{O}_{\mathrm{M}}(\mathrm{D} + \mathrm{D}') = \mathcal{O}_{\mathrm{M}}(\mathrm{D}) \otimes \mathcal{O}_{\mathrm{M}}(\mathrm{D}').$$

One defines in an obvious way the *sum* $\mathscr{G} + \mathscr{H}$ of two sub-$\mathcal{O}_{\mathrm{M}}$-Modules of an $\mathcal{O}_{\mathrm{M}}$-Module $\mathscr{F}$, as well as the product $f\mathscr{F}$ of an $\mathcal{O}_{\mathrm{M}}$-Module $\mathscr{F}$ by a holomorphic function f on M.

Also, the dual $\mathscr{F}^\vee$ of an $\mathcal{O}_{\mathrm{M}}$-Module $\mathscr{F}$ can be defined: it has as fiber, at each point $x \in \mathrm{M}$, the $\mathcal{O}_x$-module $\mathscr{F}_x^\vee$, the dual of the $\mathcal{O}_x$-module $\mathscr{F}_x$. If D is a divisor, then:

$$(10) \qquad \mathcal{O}_{\mathrm{M}}(-\mathrm{D}) = \mathcal{O}_{\mathrm{M}}(\mathrm{D})^\vee.$$

Finally, a homomorphism $u : \mathscr{F} \to \mathscr{G}$ of sheaves of groups is defined by giving, for each open set U, a group homomorphism $u_{\mathrm{U}} : \mathscr{F}(\mathrm{U}) \to \mathscr{G}(\mathrm{U})$, these homomorphisms being compatible with the restrictions. This defines, for every $x \in \mathrm{M}$, a homomorphism $u_x : \mathscr{F}_x \to \mathscr{G}_x$ on the fibers.

9. It can be said that the notion of sheaf corresponds, better than that of fiber space, to the idea of "continuous variation" of groups or of modules parametrized by the points of a holomorphic manifold. The greater generality of sheaves can be seen, for example, in the fact that if $\mathrm{N} \subset \mathrm{M}$ is a holomorphic submanifold of M and $\mathscr{F}$ an $\mathcal{O}_{\mathrm{N}}$-Module, then $\mathscr{F}$ can be "extended" to an $\mathcal{O}_{\mathrm{M}}$-Module $\mathscr{F}'$ whose fibers

$\mathscr{F}'_x$, for $x \notin N$, are *reduced to* 0. For example, if a finite set is taken for N then a "skyscraper sheaf," having only a finite number of fibers $\neq \{0\}$, is obtained. An example is given by the quotient $\mathcal{O}_{\mathbf{C}}/z\mathcal{O}_{\mathbf{C}}$ over $\mathbf{C}$, whose only nonzero fiber is the fiber at the point $0 \in \mathbf{C}$. In other words, there can be "discontinuous variations" in the fibers.

10. It can also be said that a sheaf is the mathematical expression of the idea of families of elements only *locally* defined on the manifold M; the existence of *globally* defined elements corresponds to that of *section* of a sheaf (over M). For example, a holomorphic function on M is a section of $\mathcal{O}_M$ over M, and it is known that, if M is compact, the only holomorphic functions are the constants, although over a small enough open set U, $H^0(U, \mathcal{O}_M)$ is a vector space of infinite dimension. The importance of the *cohomology groups with values in a sheaf of commutative groups* is that they evaluate the "obstructions," as it were, to the passage from the local to the global. They can be defined in several ways but their essential property is the *exact sequence of cohomology*: to each sheaf $\mathscr{F}$ of commutative groups on M and to each integer $j \geqslant 1$ is associated a cohomology group $H^j(M, \mathscr{F})$ (also denoted $H^j(\mathscr{F})$) such that for all exact sequences $0 \to \mathscr{N} \to \mathscr{G} \to \mathscr{G}/\mathscr{N} \to 0$ of sheaves of commutative groups on M, there is an exact sequence:

$$(11) \qquad \begin{aligned} 0 \to H^0(\mathscr{N}) \to H^0(\mathscr{G}) \to H^0(\mathscr{G}/\mathscr{N}) \to H^1(\mathscr{N}) \\ \to H^1(\mathscr{G}) \to H^1(\mathscr{G}/\mathscr{N}) \to H^2(\mathscr{N}) \to H^2(\mathscr{G}) \to \cdots \end{aligned}$$

We note also that the group $\mathrm{Pic}(M) = \mathbf{G}/\mathbf{G}_l$ of *divisor classes* on M (VII, 52) can be expressed in terms of the cohomology groups: it is easily seen that Pic(M) is isomorphic to $H^1(M, \mathcal{O}^*_M)$, where $\mathcal{O}^*_M$ is the sheaf of multiplicative groups such that the sections $s \in H^0(U, \mathcal{O}^*_M)$ are the *holomorphic functions nonzero everywhere on U.*

11. In the course of the year 1953, the work of Dolbeault, Serre, Kodaira, Spencer, and Hirzebruch led to a whole series of remarkable results linking Hodge theory, the invariants of the Italian geometry, and sheaf cohomology, and culminating in a formulation of the Riemann-Roch theorem (in the form of an *equality*) valid in all dimensions.

First of all, Dolbeault showed that, on a connected compact Kähler manifold M, there is a canonical isomorphism:

$$(12) \qquad \mathbf{H}^{r,s} \simeq H^s(\Omega^r_M),$$

with the preceding notation and also that of (VII, 7).

In addition, Serre discovered that the symmetric role played by the divisors D and $\Delta - D$ in the Riemann-Roch theorem for curves (VI, 14) came from a general property of *duality* that could be formulated generally for every *compact holomorphic manifold* M (not necessarily Kähler) of (complex) dimension n: for every divisor D on M, the $H^j(\mathcal{O}_M(D))$ are finite dimensional vector spaces and there is a natural duality between the spaces

$$(13) \qquad H^j(\mathcal{O}_M(D))$$

and

$$H^{n-j}(\Omega_M^n \otimes \mathcal{O}_M(-D)) = H^{n-j}(\mathcal{O}_M(\Delta - D)).$$

Severi's relation (8), proved by Kodaira, also can be written:

(14) $p_a = \dim H^0(\Omega_M^n) - \dim H^0(\Omega_M^{n-1}) + \cdots + (-1)^{n-1} \dim H^0(\Omega'_M)$,

and it follows from (12) and from the isomorphism $\mathbf{H}^{r,s} \simeq \mathbf{H}^{s,r}$ (VII, 7) that this equality can be written yet again:

(15) $p_a = \dim H^n(\mathcal{O}_M) - \dim H^{n-1}(\mathcal{O}_M) + \cdots + (-1)^{n-1} \dim H^1(\mathcal{O}_M)$.

12. The formula (15) and the analogy with the formation of the Euler-Poincaré characteristic (VIII, 5) led Serre, in 1953, to introduce for each divisor D, on a connected, compact Kähler manifold M of (complex) dimension n, the "Euler-Poincaré characteristic of D":

(16)
$\chi(D) = \dim H^0(\mathcal{O}_M(D)) - \dim H^1(\mathcal{O}_M(D)) + \cdots + (-1)^n \dim H^n(\mathcal{O}_M(D))$

such that if M is an algebraic variety, then:

$$\chi((1)) = 1 + (-1)^n p_a.$$

Taking the duality (13) into account, the Riemann-Roch theorem for a curve M can be written:

$$\chi(D) = \deg(D) - g + 1,$$

which can also be put in the form:

(17) $$\chi(D) = \left\langle f + \frac{1}{2}c_1, M \right\rangle,$$

where f is the first Chern class of the bundle $\mathbf{B}(D)$ and c_1, the first Chern class of M (here $\langle c_1, M \rangle$ is the Euler-Poincaré characteristic $2 - 2g$). Inspired by Kodaira's methods, Serre next obtained, for M a compact Kähler surface and D a divisor whose components are nonsingular curves, the formula:

(18) $$\chi(D) = p_a + 1 + \frac{1}{2}(D.(D - \Delta))$$

and, by using the duality theorem (13) that gave him

$$\dim H^2(\mathcal{O}_M(D)) = \dim H^0(\mathcal{O}_M(\Delta - D)) = l(\Delta - D),$$

he arrived at the interpretation of the "superabundance" of D (VIII, 6) as $\dim H^1(\mathcal{O}_M(D))$. In addition, using Hodge's index theorem (VII, 8) he obtained for p_a the expression:

(19) $$p_a + 1 = \left\langle \frac{1}{12}(c_2 + c_1^2), M \right\rangle$$

and the expression of (18) analogous to that of (17):

(20) $$\chi(D) = \left\langle \frac{1}{2}f(f + c_1) + \frac{1}{12}(c_2 + c_1^2), M \right\rangle.$$

We note that, when $D = (1)$, the preceding interpretation of the "superabundance" gives the inequality $q = \frac{1}{2}R_1$ (VI, 48), since $\dim H^1(\mathcal{O}_M) = p_g - p_a = q$ by (15) and $H^1(\mathcal{O}_M)$ is isomorphic to $\mathbf{H}^{1,0}$ by (12) and duality.

Further, for the canonical divisor Δ, it follows from (13) that:

$$\chi(\Delta) = p_a + (-1)^n, \tag{21}$$

a relation that is equivalent to a conjecture of Severi.

13. From this time, taking duality into account, Serre conjectured that the general form of the Riemann-Roch theorem had to give an expression of $\chi(D)$ as a function of the Chern class f of $\mathbf{B}(D)$ and the Chern classes $c_1, c_2, \ldots, c_n$ of the manifold M. He published neither the formula (20) nor this general conjecture but the latter was communicated to Kodaira and Spencer who, for their part, were working, at this time, on a conjecture originating in the work of J. A. Todd dating from 1937. J. A. Todd, following earlier research of Severi on the equivalence of cycles of arbitrary dimension, had introduced "canonical classes" of algebraic cycles of (complex) dimension $n-j$ on a nonsingular projective algebraic variety (these classes were simultaneously discovered by Eger); and he had conjectured that the arithmetic genus of M could be expressed by universal formulas using these canonical classes, formulas that he had given explicitly for $n \leqslant 6$. Later, Vesentini and others recognized that the canonical classes of Eger-Todd correspond exactly, by Poincaré duality, to the Chern classes of M. The general form of Todd's conjecture has the elegant expression, due to Hirzebruch:

$$(-1)^n p_a + 1 = \langle T_n(c_1, \ldots, c_n), M\rangle, \tag{22}$$

where T_n is a polynomial with rational coefficients, called the "Todd polynomial," which is obtained by the following calculation. In the power series:

$$\prod_{j=1}^{n} \frac{\gamma_j z}{1 - \exp(\gamma_j z)} \tag{23}$$

the coefficient of z^n, which is a symmetric polynomial in the variables γ_j, is considered. It is expressed using the elementary symmetric functions σ_j of the γ_j $(1 \leqslant j \leqslant n)$; then, in the polynomial obtained, σ_j is replaced by c_j. For example, the first Todd polynomials are:

$$T_1(c_1) = \frac{1}{2}c_1, \quad T_2(c_1, c_2) = \frac{1}{12}(c_1^2 + c_2),$$

$$T_3(c_1, c_2, c_3) = \frac{1}{24}c_1 c_2$$

$$T_4(c_1, c_2, c_3, c_4) = \frac{1}{720}(-c_4 + c_3 c_1 + 3c_2^2 + 4c_2 c_1^2 - c_1^4).$$

Todd had been able to prove (22) only by relying on an unproven conjecture of Severi. Toward the end of 1953, guided by Serre's results and using Thom's results on cobordism, Hirzebruch succeeded in proving not just relation (22) but the general Riemann-Roch theorem in the form conjectured by Serre:

(24) $$\chi(\mathrm{D}) = \langle \mathrm{P}(f, c_1, \ldots, c_n), \mathrm{M} \rangle$$

where P is a polynomial obtained by the same procedure as the Todd polynomial, but starting from the power series:

$$e^{fz} \prod_{j=1}^{n} \frac{\gamma_j z}{1 - \exp(\gamma_j z)}.$$

14. The initial "Riemann-Roch problem," that is to say, the actual calculation of $l(\mathrm{D})$, is resolved by formula (24) only if $\chi(\mathrm{D}) = l(\mathrm{D})$, in particular, if $\mathrm{H}^j(\mathcal{O}_{\mathrm{M}}(\mathrm{D})) = 0$ for $j \geqslant 1$. In 1953, Kodaira gave an important criterion for this to be so. He considers a compact Kähler manifold M; designating by $\mathrm{H}^{1,1}(\mathrm{M}, \mathbf{R})$ the subspace of $\mathrm{H}^1(\mathrm{M}, \mathbf{R})$ that corresponds to the subspace $\mathbf{H}^{1,1}_{\mathbf{R}}$ of real harmonic 2-forms of type (1, 1) (VII, 7), he says that a cohomology class $\xi \in \mathrm{H}^{1,1}(\mathrm{M}, \mathbf{R})$ is *positive* if it is the class of a closed form Θ corresponding to a Kähler metric on M (not necessarily identical to the given metric) (VII, 6).

On the other hand, consider the exact sequence of sheaves of groups on M:

(25) $$0 \longrightarrow \mathbf{Z} \longrightarrow \mathcal{O}_{\mathrm{M}} \xrightarrow{\exp} \mathcal{O}^*_{\mathrm{M}} \longrightarrow 0,$$

where $\mathbf{Z}$ is the constant sheaf, each fiber of which is the additive group $\mathbf{Z}$, and exp, the exponential map that associates to a section $s \in \mathrm{H}^0(\mathrm{U}, \mathcal{O}_{\mathrm{M}})$, the section e^s of $\mathcal{O}^*_{\mathrm{M}}$ over U. From it, by (11), is deduced the exact sequence:

(26) $$\mathrm{H}^1(\mathrm{M}, \mathcal{O}^*_{\mathrm{M}}) \to \mathrm{H}^2(\mathrm{M}, \mathbf{Z}) \to \mathrm{H}^2(\mathcal{O}_{\mathrm{M}}).$$

It is easily seen that the first arrow maps the class of a divisor D (VIII, 10) to the Chern class $c_1(\mathbf{B}(\mathrm{D}))$. Kodaira and Spencer showed that the resulting image of $\mathrm{H}^1(\mathrm{M}, \mathcal{O}^*_{\mathrm{M}})$ is the group:

(27) $$\mathrm{H}^2(\mathrm{M}, \mathbf{Z}) \cap \mathrm{H}^{1,1}(\mathrm{M}, \mathbf{R}).$$

Then Kodaira deduced from this, using a method from differential geometry due to Bochner, that if a divisor D on M is such that the class $c_1(\mathbf{B}(\mathrm{D} - \Delta))$ is *positive*, then $\mathrm{H}^j(\mathcal{O}_{\mathrm{M}}(\mathrm{D})) = 0$ for all $j \geqslant 1$. In particular, if M is a projective algebraic variety and H a hyperplane section of M, then:

(28) $$\mathrm{H}^j(\mathcal{O}_{\mathrm{M}}(\mathrm{D} + m\mathrm{H})) = 0 \qquad \text{for } j \geqslant 1 \qquad \text{and} \qquad m \geqslant m_0$$

(m_0, an integer independent of j).

Furthermore, we note that the exact sequence (26) allows the generalization, to algebraic varieties of arbitrary dimension n, of Lefschetz's result on the characterization of the 2-cycles of a surface homologous to a divisor (VII, 53): a cohomology class of $\mathrm{H}^2(\mathrm{M}, \mathbf{Z})$ is the class of a divisor if its canonical image in $\mathrm{H}^2(\mathcal{O}_{\mathrm{M}})$ is zero.

15. Using his criterion for the vanishing of the higher cohomology groups, Kodaira finally was able to prove his fundamental criterion for a compact Kähler

manifold to be isomorphic to a projective algebraic variety: it is necessary and sufficient that there exists a Kähler metric on M such that the class of the corresponding 2-form Θ (VII, 6) belongs to $H^2(M, \mathbf{Z})$. In that case, this class is the image of the class of a divisor D. Inspired by a method of Castelnuovo-Enriques using suitable blowups of M, Kodaira succeeded in proving, by applying the exact cohomology sequence and his vanishing criterion, that a large enough multiple of D allows the definition of an embedding of M in a $\mathbf{P}_N(\mathbf{C})$ (VI, 58). Another proof of Kodaira's criterion was given by Grauert.

16. We point out how to translate the definitions and principal results of the Italian geometers into sheaf cohomology. The plurigenera of Enriques are given by:

(29) $$P_k = \dim H^0(\mathcal{O}_M(k\Delta)).$$

When the divisor Δ is positive, it follows from the Riemann-Roch theorem (24) that, for k tending toward $+\infty$:

(30) $$P_k \sim \langle(-c_1)^n, M\rangle . k^n/n! \qquad (n = \dim(M)).$$

The linear genus of a surface M (VI, 47) is related to the first Chern class by the formula:

(31) $$p^{(1)} = \langle c_1^2, M\rangle + 1.$$

Furthermore, if a linear system of irreducible curves of dimension 1 on a surface M is considered, then Zeuthen and Segre had discovered that the number $I = \delta - n - 4\pi$ is an invariant of the surface M where n is the degree, π the genus of a generic curve of this system, and δ the number of curves of the system that are of genus $\pi - 1$. This invariant of Zeuthen-Segre is found to be equal to $\chi(M) - 4 = \langle c_2, M\rangle - 4$ (Picard-Alexander), and, together with (31), the formula (21) for the arithmetic genus of M gives again a formula already discovered by M. Noether:

(32) $$p^{(1)} + I = 12p_a + 9.$$

17. By studying the arithmetic genus of a surface and the Riemann-Roch problem, the Italian geometers had examined in detail the situation where an irreducible curve C and a divisor D on a nonsingular irreducible projective surface M are given and where the *linear series* $|D|.C$ on the curve C formed by the divisors $D'.C$, where D' ranges over the set of the divisors of the complete system $|D|$, is considered. In general, this series is not complete, and the difference between the dimension of the complete series $|D.C|$ on C and the series formed of the $D'.C$ is called the *deficiency* of $|D|$ on C. They had obtained three interesting results:

1) If C_m is the intersection of M and a "general" surface of degree m, the deficiency of $|D|$ on C_m is zero for large enough m (lemma of Enriques-Severi).

2) When $D = C + \Delta$, the system $|D|$ is called the *adjoint* of $|C|$; then the deficiency of $|D|$ on C is at most equal to the *irregularity* $q = p_g - p_a$ of M and can

attain this value (a theorem of Enriques); moreover, Picard and Severi proved that this maximum q is always attained if $|C|$ has dimension $\geqslant 2$; in the case where the maximum is attained, $l(\Delta - C) = 0$ and the "superabundance" of $|C|$ (VIII, 6) is zero.

3) If $D = C$, the deficiency of $|C|$ on C is again at most equal to q (a theorem of Castelnuovo); the linear series $|C|.C$ on C is called the "characteristic series" of C.

18. Zariski and Kodaira generalized these results to projective varieties of arbitrary dimension. If S is an irreducible hypersurface in a projective variety M and D a divisor on M, the vector bundle $\mathbf{E} = \mathbf{B}(D)$ and its restriction $\mathbf{E}|S$ to S are considered; by restriction to S, a canonical homomorphism $r_s : \Gamma(\mathbf{E}) \to \Gamma(\mathbf{E}|S)$ is obtained, and the *deficiency* $\operatorname{def}(D/S)$ is the difference $\dim(\Gamma(\mathbf{E})) - \dim r_s(\Gamma(\mathbf{E}))$. With $D = D_0 + S$, there is an exact sequence of sheaves:

$$0 \to \mathcal{O}_M(D_0) \to \mathcal{O}_M(D_0 + S) \to \mathcal{O}_S((D_0 + S).S) \to 0$$

and the homomorphism r_s is the one coming from the exact cohomology sequence:

$$(33) \qquad \cdots \to H^0(\mathcal{O}_M(D_0 + S)) \xrightarrow{r_s} H^0(\mathcal{O}_S((D_0 + S).S)) \to \to H^1(\mathcal{O}_M(D_0)) \to H^1(\mathcal{O}_M(D_0 + S)) \to \cdots$$

Thus, the inequality

$$(34) \qquad \operatorname{def}((D_0 + S)/S) \leqslant \dim H^1(\mathcal{O}_M(D_0))$$

follows, with equality attained if $H^1(\mathcal{O}_M(D_0 + S)) = 0$, which is the case if the class $c_1(\mathbf{B}(D_0 + S - \Delta))$ is positive (VIII, 14). For $D_0 = \Delta$, the generalization of the theorem of Enriques with:

$$q = \dim H^1(\mathcal{O}_M(\Delta)) \qquad (= \dim H^1(\mathcal{O}_M)),$$

by Serre duality, when M is a surface), is obtained; for $D_0 = 0$, this gives the generalization of Castelnuovo's theorem (with the same value of q). Finally, if $S = S_m$, the intersection of M and a general hypersurface of degree m, the lemma of Enriques-Severi results from the relation, proved by Zariski and Kodaira:

$$(35) \qquad H^1(\mathcal{O}_M(D - S_m)) = 0 \qquad \text{for } m \text{ large enough.}$$

2. The varieties of Serre

19. Around 1949, A. Weil had observed that the "Zariski topology" (VII, 45) could be defined on his "abstract varieties" (VII, 38) as well, and not only did it simplify the exposition but it made possible the definition in "abstract" algebraic geometry of the notion of *fiber space*, modeled on the classical notion (VIII, 2), as well as the extension of the relations between line bundles and divisors (VIII, 4). In 1954, Serre had the idea of similarly generalizing the notion of *sheaf* to "abstract" varieties by replacing, in the definition of (VIII, 8), the usual topology by the Zariski topology. But, owing to this idea, he realized that the definition and

the study of Weil's "abstract varieties" could even be presented in a much more convenient form, using H. Cartan's notion of "ringed space," that is to say, a topological space X on which is given a *sheaf of rings*, called the *structure* sheaf (that is, a sheaf where, for each open set $U \subset X$, $H^0(U, \mathcal{O}_X)$ has the structure of a ring such that the restriction homomorphisms $H^0(U, \mathcal{O}_X) \to H^0(V, \mathcal{O}_X)$ for $V \subset U$ are ring homomorphisms). The advantage of this point of view is that this type of structure lends itself very well to "gluing" along open sets, the verification of the "conditions for gluing" ordinarily being trivial.

20. For the applications he has in mind, Serre is not concerned with questions of the "field of definition" (VII, 36) and he does not use the notion of "generic point"; staying closer to the classical point of view, he fixes, once and for all, an *algebraically closed* base field k (of arbitrary characteristic). The "pieces" that he "glues together" to obtain the definition of his varieties are what are called *affine varieties* over k; such a variety X is a subset of a space k^n defined by polynomial equations, and the sheaf $\mathcal{O}_X$ is defined by the condition that, for every open set $U \subset X$, $H^0(U, \mathcal{O}_X)$ is composed of the restrictions to U of the rational functions $x \mapsto P(x)/Q(x)$ on k^n that are defined (that is, such that $Q(x) \neq 0$) at every point $x \in U$. In addition, Serre imposed on his varieties, on the one hand, the condition of having a covering by a finite number of affine open sets (that is, open sets U such that the restriction to U of the sheaf $\mathcal{O}_X$ defines U as an affine variety), and on the other, a "separation condition" allowing "passage to the limit" for the Zariski topology in a reasonable manner: for example, if two rational functions f, g are defined in X and agree on an everywhere dense open set of X, then they are necessarily identical.

21. Serre's principal goal is to extend, as much as possible, to his varieties the results on sheaf cohomology described above for the classical case ($k = \mathbf{C}$). He restricts to *coherent* $\mathcal{O}_X$-Modules (in order to be able to use the exact cohomology sequence (formula (11)) with the definition of the cohomology groups ("Čech cohomology") of which he avails himself). Coherent $\mathcal{O}_X$-Modules can be defined in the following way. First of all, if X is an affine variety and:

$$A = \Gamma(X, \mathcal{O}_X)$$

is the (noetherian) ring of regular functions on X (VIII, 27), let M be an A-module of finite type; for each function $f \in A$, the set $D(f)$ of the $x \in X$ such that $f(x) \neq 0$ is a Zariski open set; the coherent sheaf $\mathcal{F}$ corresponding to M is such that $H^0(D(f), \mathcal{F})$ is the module composed of the elements m/f^k (k an arbitrary integer, $m \in M$). When X is arbitrary, the restriction of $\mathcal{F}$ to each affine open set must be of the preceding form.

Serre is interested principally in closed subvarieties X of a projective space $\mathbf{P}_N(k)$; in this case, he succeeds in replacing Kodaira's analytic methods by a purely algebraic technique, that of canonically associating to a coherent sheaf $\mathcal{F}$ a graded module over the ring $k[T_0, T_1, \ldots, T_N]$. Generalizing the theorem on the finiteness of $l(D)$, he shows, first, that for a coherent $\mathcal{O}_X$-Module $\mathcal{F}$, the groups

$H^q(\mathscr{F})$ are vector spaces of *finite* dimension over k, and zero for $q > \dim(X)$. He also generalizes the formation of the divisors $D + nH$ (n a positive or negative integer, H a hyperplane section) from the Italian geometry by forming for every coherent $\mathcal{O}_X$-Module $\mathscr{F}$, the $\mathcal{O}_X$-Modules $\mathscr{F}(n) = \mathscr{F} \otimes_{\mathcal{O}_X} \mathcal{O}_X(nH)$. Then, he extends Kodaira's result (28) by showing that there exists an integer n_0 such that, for $n \geqslant n_0$:

$$H^q(\mathscr{F}(n)) = 0 \qquad \textit{for all} \qquad q \geqslant 1.$$

Finally, he discovers that the cohomology of coherent sheaves is closely connected to the Ext functors from homological algebra; for example, in order that $H^q(\mathscr{F}(-n)) = 0$ for n sufficiently large it is necessary and sufficient that:

$$\mathrm{Ext}_{A_x}^{N-q}(\mathscr{F}_x, A_x) = 0 \qquad \text{for all} \qquad x \in X,$$

where A_x is the local ring of $\mathbf{P}_N(k)$ at the point x. By assuming X *normal*, he deduces from this, the Enriques-Severi lemma in the form (35) (it must be emphasized that *no* restriction was imposed on the singular points of X for the preceding properties).

22. Now, by assuming that $X \subset \mathbf{P}_N(k)$ is *nonsingular*, Serre shows that his duality theorem is still valid in the more general framework where it is placed. Of course, it is necessary to define the sheaves Ω^r_X of germs of differential r-forms in a purely algebraic way; evidently, that amounts to defining the notion of "tangent vector" since the definition of the bundle of the r-cotangent vectors can be deduced algebraically in the usual way. As the question is local, the varieties may be assumed to be affine, and, since the varieties under consideration have all points simple, by the jacobian criterion and the fact that the field k is perfect (VII, 41), the classical definition may simply be transcribed. Then, the duality theorem extends in the following general form: if $\mathbf{E}$ is a vector bundle with base a projective variety X, $\check{\mathbf{E}}$ its dual, and $\mathbf{T}^r$ the bundle of r-cotangent vectors (so that $\mathcal{O}(\mathbf{T}^r) = \Omega^r_X$), the k-vector spaces $H^q(\mathcal{O}(\mathbf{E}))$ and $H^{n-q}(\mathcal{O}(\check{\mathbf{E}} \otimes \mathbf{T}^n))$ are canonically in duality (with $n = \dim(X)$).

23. Generalizing formula (16), Serre also defines the "Euler-Poincaré characteristic" of an arbitrary coherent sheaf $\mathscr{F}$ on X:

(36)
$$\chi(\mathscr{F}) = \dim H^0(\mathscr{F}) - \dim H^1(\mathscr{F}) + \cdots + (-1)^n \dim H^n(\mathscr{F}) \qquad (n = \dim(X)).$$

He shows that $\chi(\mathscr{F}(m))$ is equal to a polynomial in m for *all* values of $m \in \mathbf{Z}$; as $H^q(\mathscr{F}(m)) = 0$ for $q \geqslant 1$ and m sufficiently large, by taking $\mathscr{F} = \mathcal{O}_X$, this recovers the Hilbert polynomial (VII, 17) and "explains" its cohomological signification for *any m* whatsoever.

From that moment, the question arose of generalizing Hirzebruch's expression (24) for coherent sheaves on a nonsingular projective variety (over an algebraically closed base field of arbitrary characteristic). This is what Grothendieck and Washnitzer independently did in 1957. It must be pointed out that Grothendieck deduces this formula from another more general one, that refers no longer to a variety X but to a morphism $f: Y \to X$ (VIII, 25) and to the relations between the Chern classes of a coherent $\mathcal{O}_X$-Module $\mathcal{F}$ and its inverse image under f, which is a coherent $\mathcal{O}_Y$-Module. It is for this reason that he introduced the idea of "K-theory" that was, in the years following, to invade not only differential topology but even algebra. We recall that, at present, Hirzebruch's formula (24) is deduced as the particular case $k = \mathbf{C}$ of a formula from the K-theory of differential manifolds, the Atiyah-Singer formula.

24. In the classical case ($k = \mathbf{C}$) there are two topologies on a projective variety M, the Zariski topology and the finer "usual" topology; the structure on $\mathbf{P}_N(\mathbf{C})$ of holomorphic manifold permits also the definition on M of a structure of "analytic space" (having "singularities," in general) and of "analytic sheaves." To every $\mathcal{O}_M$-Module $\mathcal{F}$ is canonically associated the analytic sheaf $\mathcal{F}^h = \mathcal{F} \otimes_{\mathcal{O}_M} \mathcal{H}$, where $\mathcal{H}$ is the sheaf of germs of holomorphic functions on M. In 1956, Serre made a detailed study of the relations between these structures and showed, among other things, that the map $\mathcal{F} \mapsto \mathcal{F}^h$ defines a one-to-one correspondence between coherent algebraic sheaves and coherent analytic sheaves, and gives an isomorphism of cohomology groups, allowing the interchangeable use of algebraic methods or the analytic methods of Hodge-Kodaira.

25. Being principally interested in questions of cohomology, Serre did not develop the general properties of his varieties in detail. That was done, almost at the same time, by Chevalley, using a definition of variety different from that of Serre and less tractable, but equivalent. One of the principal points to underline is that with Serre and more so with Chevalley, birational geometry distinctly fades into the background and the concept of *morphism* becomes the fundamental idea. Up until then, the interest was concentrated on *complete* varieties (of which projective varieties are particular examples) and it is rare that a rational correspondence from X to Y, when X and Y are complete, is defined at *every* point of X. On the contrary, if X and Y are two arbitrary Serre varieties, a morphism of $f: X \to Y$ is, first of all, a *map* of the (whole) set X into Y, which must be continuous and transform each section $s \in H^0(V, \mathcal{O}_Y)$ into a section $s \circ f \in H^0(f^{-1}(V), \mathcal{O}_X)$ for every open set $V \subset Y$.

26. The varieties of Serre form the natural framework of the theory of *linear algebraic groups* on an algebraically closed field of *arbitrary* characteristic, which was developed by C. Chevalley and A. Borel, between 1951 and 1957, in a series of remarkable memoirs. Their joint efforts succeeded in modeling this theory on the theory of Lie groups by employing only *global* techniques, the peculiarities coming from characteristic $p > 0$ prohibiting the "lifting" of the Lie algebras to the global groups as in the classical theory. The climax of this work was the complete classification, due to Chevalley, of the *semi-simple* algebraic groups over an algebraically closed field of characteristic $p > 0$. It is (up to questions of "isogeny") *identical* to the classification of Killing–E. Cartan, contrary to what would *à priori* be supposed because of the existence of numerous "nonclassical" simple Lie algebras over a field of characteristic $p > 0$.

3. Schemes and Topologies

27. Up until 1950, no one seems to have tried to give an *intrinsic* definition of an affine algebraic variety over an *algebraically closed field* k, which is independent of the embedding of the variety in an affine space k^n. Nevertheless, it was very easy to do it using only Hilbert's theorem of zeros (VII, 16), one form of which is that each maximal ideal of the algebra of polynomials $k[T_1, \ldots, T_n]$ is generated by n polynomials of the first degree $T_1 - \zeta_1, \ldots, T_n - \zeta_n$ with:

$$(\zeta_1, \ldots, \zeta_n) \in k^n.$$

Just as Riemann had attached to a projective algebraic curve, the field of rational functions on this curve (V, 14), so to an affine variety $V \subset k^n$ can be attached the ring $A(V)$ of the restrictions to V of all the polynomial functions on k^n. This ring is a k-algebra of finite type that is *reduced* (in other words, has no nonzero nilpotent elements); the theorem of zeros shows that the points of V are canonically in one-to-one correspondence with the *maximal ideals* of $A(V)$. *Conversely*, it is easily seen that *every* reduced k-algebra of finite type is of the form $A(V)$ for an affine variety determined up to isomorphism. Moreover, when V is irreducible, it is even possible to define the structure sheaf $\mathcal{O}_V$ directly from the ring $A(V)$: for every (Zariski open) set $U = D(f)$ of V, defined as the set of the $x \in V$ such that $f(x) \neq 0$ for an $f \in A(V)$, $\mathcal{O}_V(U)$ is defined as the ring of rational functions of the form g/f^m for $g \in A(V)$ and m an integer > 0. It is easy to see that this completely defines $\mathcal{O}_V$. Finally, if V, W are two affine varieties, to each morphism $u : V \to W$ corresponds the homomorphism of k-algebras $A(W) \to A(V)$ that associates to each function $g \in A(W)$, the function $g \circ u$. Conversely, for *every* k-homomorphism of algebras $\phi : A(W) \to A(V)$, the inverse image $\phi^{-1}(\mathfrak{m})$ of a maximal ideal $\mathfrak{m}$ of $A(V)$ is a maximal ideal of $A(W)$, and it is proved that $\mathfrak{m} \mapsto \phi^{-1}(\mathfrak{m})$ is a morphism to which ϕ corresponds by the preceding procedure. In the language of categories, the use of which began to spread around 1955, the category of affine varieties over k is *equivalent* to the *dual* category of the category of (commutative) reduced k-algebras of finite type.

28. Following P. Cartier's suggestion, A. Grothendieck undertook, around 1957, a gigantic program whose objective was a vast generalization of algebraic geometry, absorbing all the earlier developments, and starting from the category of *all* commutative rings in place of the subcategory of the reduced algebras of finite type over an algebraically closed field.

However, to define, as above, a category that is equivalent to the dual of the category of all commutative rings, the method of (VIII, 27) must be appreciably modified from the start: if $\phi : A \to B$ is a ring homomorphism, the inverse image $\phi^{-1}(\mathfrak{m})$ of a maximal ideal $\mathfrak{m}$ of B is not, in general, a maximal ideal of A; by contrast, for each *prime* ideal $\mathfrak{p}$ of B, $\phi^{-1}(\mathfrak{p})$ is always a prime ideal of A. Thus, it is necessary in order to replace the "affine variety," to consider the *spectrum* of A, that is, the set Spec(A) of all the prime ideals of A; a "Zariski topology" is defined on this set by taking the sets $V(\mathfrak{a})$ for closed sets, where $V(\mathfrak{a})$ is the set of prime ideals containing an (arbitrary) ideal $\mathfrak{a}$; it must be noted that, in general, finite sets are no longer closed for this topology. Finally, by using the general definitions of localization, given by Chevalley and Uzkov in 1943–1948, it is possible to give meaning to g/f^m even when f is a divisor of zero in A, and, consequently, the structure sheaf $\mathcal{O}_X$ on $X = \mathrm{Spec}(A)$ can be defined in the same way as for affine varieties. The ringed spaces (VIII, 19) thus obtained are called *affine schemes*, and form a category equivalent to the dual of the category of all commutative rings. From there, the category of *schemes* is reached by the process of "gluing" similar to that which defined the varieties of Serre (VIII, 20) but starting from affine schemes instead of affine varieties, and imposing *no* restriction on the affine open sets of the scheme and no "separation condition."

29. The experience of the last 25 years has convinced the specialists that, in spite of the much more advanced techniques of commutative algebra that it necessitates, the theory of schemes is the context in which the problems of algebraic geometry are best approached and understood.

Clearly, the theory of schemes includes, by definition, all of commutative algebra as well as all of the theory of the varieties of Serre. Thus, a part of the theory consists in transcribing, in a more or less evident way, the results of these more specialized theories as far as they are still valid. In particular, we note that the notion of cohomology groups $H^q(\mathcal{F})$ of an $\mathcal{O}_X$-Module $\mathcal{F}$ can be defined on a scheme X, having the fundamental property of the existence of the exact sequence of cohomology (VIII, 10), without imposing any supplementary condition on $\mathcal{F}$. This is due to a new definition of these groups that was introduced by Grothendieck by applying the theory of abelian categories, and that is more flexible than Čech cohomology, although less convenient in calculations. The sheaves used most often are the *quasi-coherent* $\mathcal{O}_X$-Modules $\mathcal{F}$, the definition of which is the same as that of the coherent $\mathcal{O}_X$-Modules (VIII, 21) except that no condition is imposed on the module M. An important technical result is Serre's affineness criterion: if X is an affine scheme, then $H^q(\mathcal{F}) = 0$ for every quasi-coherent $\mathcal{O}_X$-Module $\mathcal{F}$ and all $q \geqslant 1$; conversely, if $H^1(\mathcal{F}) = 0$ for every quasi-coherent $\mathcal{O}_X$-Module $\mathcal{F}$, then X is an affine scheme.

30. We will pause more particularly on the properties of schemes that have no analogue in the earlier theories. In the first place, the notion of *specialization* is, in this setting, of a topological nature: a point x' of a scheme X is a specialization of a point x if x' belongs to the closure $\overline{\{x\}}$. In particular, a *generic point* is, now, a point ξ of X whose closure is all of X; for example, if A is an integral domain, the unique generic point of Spec(A) is the prime ideal (0). Thus, continuity arguments (of course, for the Zariski topology) can again be used, as the Italian geometers did, and the multitude of "generic points," that bothered Zariski in Weil's conception (VII, 36), can be eliminated.

31. In the theory of schemes, the accent is put on the properties of *morphisms* more than on the properties of the schemes themselves. Most often, the object of

study is a morphism of schemes $u : X \to S$, where S is often arbitrary (for *fixed* S, it is said that the study of these morphisms is the study of "S-schemes"). This point of view is particularly apparent when one has to impose *finiteness conditions* (without which there is little hope of obtaining significant results). Indeed, Grothendieck has shown that, except for cohomological notions, *no* finiteness condition (such as the fact of being noetherian, or of finite dimension, etc.) need, in general, be imposed on the "base scheme" S, provided that finiteness conditions (finite type, finite presentation, etc.) are imposed on the morphism u; this gives considerable flexibility in the handling of "base changes" (see below).

An aspect of this "relativization" of the fundamental notions is the concept of "higher images" of a sheaf $\mathscr{F}$ on a scheme X under a morphism $u : X \to Y$, which "relativizes" the notion of cohomology group: assuming, for simplicity, that u is of finite type and that $\mathscr{F}$ is a quasi-coherent $\mathcal{O}_X$-Module, the higher image $R^q u_* (\mathscr{F})$ of $\mathscr{F}$ is the $\mathcal{O}_Y$-Module such that, for each affine open set U of Y, the group of sections of this Module over U is the cohomology group $H^q(\mathscr{F}|u^{-1}(U))$ of the restriction of $\mathscr{F}$ to the open set $u^{-1}(U)$.

32. Given two "S-schemes" $f : X \to S$, $g : Y \to S$, there is a triple, unique up to isomorphism, composed of an S-scheme $X \times_S Y$ and two morphisms $p_1 : X \times_S Y \to X$, $p_2 : X \times_S Y \to Y$, such that

$$f \circ p_1 = g \circ p_2,$$

and that is the "product," in the categorical sense, of X and Y over S. This means that if there are two morphisms $u : T \to X$, $v : T \to Y$ such that $f \circ u = g \circ v$, then there is a unique morphism $w : T \to X \times_S Y$ such that $u = p_1 \circ w$ and $v = p_2 \circ w$. This follows easily from the existence of a tensor product $B \otimes_A C$ for two arbitrary A-algebras B, C over any ring A. It is worth noting that there is no analogous operation for the varieties of Serre.

This construction is most often applied to study the morphism $f : X \to S$ by replacing the "base" S by another scheme S′ equipped with a morphism $S' \to S$; then $X_{(S')}$ is written in place of $X \times_S S'$ and $f_{(S')} : X_{(S')} \to S'$ in place of p_2, and $X_{(S')}$ and $f_{(S')}$ are said to have been obtained from X and f by *base change*. This is, without doubt, the most powerful tool of the theory of schemes. It has an extraordinary flexibility, that generalizes, in numerous unforeseen ways, the old idea of "extension of scalars." We will give only a few examples.

33. First of all, at every point s of a scheme S, there is, as in the case of classical varieties, a *local ring* $\mathcal{O}_s$, the fiber of the sheaf $\mathcal{O}_S$ at this point; if $\mathfrak{m}_s$ is its unique maximal ideal, $\kappa(s) = \mathcal{O}_s/\mathfrak{m}_s$ is the residue field of S at the point s. Contrary to what happens for the varieties of Serre (where the residue field is k at every point), the field $\kappa(s)$ varies with s (it is essentially the "field of rational functions" on a "subscheme" of S having the closure $\overline{\{s\}}$ for underlying space so that s is its "generic point"; cf. (VII, 39)). Then, it can be verified that the scheme $X_s = X \times_S \operatorname{Spec}(\kappa(s))$, obtained by the base change $\operatorname{Spec}(\kappa(s)) \to S$, has the "fiber" $f^{-1}(s)$ in X (with the induced topology) as underlying space and that, by

means of finiteness conditions, this scheme can be considered as a "variety" over the field $\kappa(s)$ in a sense a little more general than Serre's.

34. Thus, an "S-scheme" X can be considered as a "family of varieties »X_s« parametrized" by S, generalizing, in short, Picard's idea for the study of surfaces (VI, 35). To pursue this idea, it is natural to try to deduce the properties of the morphism $f: X \rightarrow S$ from properties of the X_s. It turns out that this is often possible by means of a condition on f "gluing" the diverse fibers X_s in some way; this property, originating in homological algebra and introduced in the first place by Serre in local algebra under the name of "flatness," effectively avoids "abrupt variations" of X_s when s varies continuously in S, but can be expressed only in algebraic terms. For example, the morphism $x \mapsto x^2$ of k to itself (k being an algebraically closed field of characteristic $\neq 2$) is flat, whereas the morphism $(x,y) \mapsto (x^2, xy, y^2)$ of k^2 onto the cone with equation $x_1x_3 - x_2^2 = 0$ in k^3 is not flat.

The notion of flat morphism allows the definition of other types of morphisms that correspond to the notions of fibration (*smooth* morphisms) and of covering (*étale* morphisms) in the theory of differential manifolds. These are flat morphisms; in addition, an étale morphism has the property that each fiber $f^{-1}(s)$ is composed of isolated points, the residue field $\kappa(x)$ at each of which is a finite separable extension of $\kappa(s)$; a smooth morphism $f: X \rightarrow S$ has the property that, for each point $x \in X$, there exists an open neighborhood V of x and an open affine neighborhood U of $s = f(x)$ such that $f(V) \subset U$ and the morphism $V \rightarrow U$ induced by f factors into:

$$V \xrightarrow{h} \operatorname{Spec}(A[T_1, \ldots, T_r]) \xrightarrow{g} U = \operatorname{Spec}(A),$$

where h is étale (the intermediary scheme can be called the "affine space A^r" over the ring A).

35. The passage from a scheme on base S to the "punctual" scheme $\operatorname{Spec}(\kappa(s))$ destroys *ipso facto* all that takes place outside of the fiber X_s. If it is desired to study what takes place in some *neighborhood* of the fiber, it is only necessary to replace S by the "local" scheme $\operatorname{Spec}(\mathcal{O}_s)$, which, by means of suitable finiteness conditions, gives exactly the information desired. Indeed, $\mathcal{O}_s$ can be considered as the "inductive limit" of the rings A_λ of the open affine sets U_λ containing s, and Grothendieck has developed a technique allowing proof of the fact that when the inverse image under f of $\operatorname{Spec}(\mathcal{O}_s)$ (this scheme is also denoted $X \otimes_S \mathcal{O}_s$) has certain properties, then, for suitable λ, the inverse image $f^{-1}(U_\lambda) = X \otimes_S A_\lambda$ has these same properties. It must be noted that this same technique of "inductive limit" over the base ring permits, for example, the pulling back of the properties of a scheme over an *arbitrary* extension k of a field k_0 to the properties of a scheme over an extension of k_0 of *finite type*, or the properties of a scheme over an arbitrary ring to those of a scheme over a **Z**-algebra of finite type, realizing the old dream of Kronecker (VI, 2), and reducing many questions on S-schemes to the case where S is noetherian.

36. The theory of S-schemes, when $S = \mathrm{Spec}(A)$ is a local scheme, is particularly simple when A is a *complete* noetherian local ring. Whence comes the idea, when it is desired to study the more general case where A is local and noetherian, of passing by base change to its completion $\hat{A}$. But here we meet a particular case of the general problem of "descent": when, after a base change $S' \to S$, properties of the S′-scheme $X_{(S')}$ have been demonstrated, how to "descend" to the initial scheme X again? In the case of varieties defined over a field k, the passage of the properties of the points "with values in the algebraic closure $\bar{k}$ of k" to the points "with values in k" is a particular case of this problem. Grothendieck has made a penetrating study of questions of "descent" that play an important technical role in the questions of existence of schemes satisfying certain conditions, for example, the difficult problem of the *quotient* of a scheme by an equivalence relation. The two particular cases mentioned above are among the most simple, because then the morphism $S' \to S$ is *faithfully flat* (i.e. flat and surjective), and many properties remain unchanged under such a base change.

37. Another procedure defined in a satisfactory way by the theory of schemes, and to which the other notions of "variety" were ill adapted is what is called the "reduction modulo an ideal." A variety X over a field K, embedded in a projective space $\mathbf{P}_r(K)$, is given and (for simplicity) it is assumed that K is the field of fractions of a discrete valuation ring A so that the A-scheme $\mathbf{P}_r(A)$ (see below (VIII, 43)) can be considered; if $f: \mathbf{P}_r(A) \to \mathrm{Spec}(A)$ is the structure morphism, $\mathbf{P}_r(K)$ is identified with $f^{-1}(U)$, where U is the open set of Spec(A) consisting of the generic point (0) of Spec(A). It follows that there is a smallest closed A-scheme $X' = \bar{X}$ that contains X in $\mathbf{P}_r(A)$ and has the property that $X = X' \cap f^{-1}(U) = X' \otimes_A K$. If $\mathfrak{M}$ is the maximal ideal of A and $k = A/\mathfrak{M}$ the residue field, so that $a = \mathfrak{M}$ is the unique closed point of Spec(A), the complement of U, then the "reduction modulo $\mathfrak{M}$" of X is the k-variety $X_0 = X' \cap f^{-1}(a) = X' \otimes_A k$. It is clear that, without passing through the intermediary of an A-scheme, it is difficult to connect the K-variety X to the k-variety X_0 in a satisfactory way. Moreover, the modern point of view is to consider directly an A-scheme X′ and its fiber at the closed point of Spec(A) as the essential objects of study rather than a K-variety from which X′ would be deduced.

38. The idea of considering affine schemes Spec(A) even when A has nonzero *nilpotent* elements might seem dictated by a desire for generality for its own sake; in fact, it allows the phenomena of "multiple varieties" to be taken into consideration; these were either completely neglected by the classical theory or treated only obliquely by the theory of intersections. For example, consider the parabola $y^2 - x = 0$ in $\mathbf{C}^2$ and the map that projects it onto the axis Ox. In the language of schemes, two affine schemes are considered:

$$U = \mathrm{Spec}(\mathbf{C}[T_1, T_2]/(T_2^2 - T_1)), \qquad V = \mathrm{Spec}(\mathbf{C}[T_1])$$

and the morphism $p: U \to V$ that corresponds to the natural injection

$\mathbf{C}[T_1] \to \mathbf{C}[T_1, T_2]/(T_2^2 - T_1)$. A maximal ideal $(T_1 - \zeta)$ of $\mathbf{C}[T_1]$ is identified with the point $\zeta \in \mathbf{C}$ and the fiber $U_\zeta = p^{-1}(\zeta)$ is the affine scheme:

$$\operatorname{Spec}(\mathbf{C}[T_2]/(T_2^2 - \zeta)).$$

If $\zeta \neq 0$, the ring $\mathbf{C}[T_2]/(T_2^2 - \zeta)$ is isomorphic to a direct sum of two fields, each isomorphic to $\mathbf{C}$, corresponding to the fact that the fiber has two distinct points; but if $\zeta = 0$, $\mathbf{C}[T_2]/(T_2^2)$ has nilpotent elements, corresponding to the fact that the two points have become "infinitely near."

In general, it is in the nilpotent elements that the theory of schemes finds the algebraic equivalent of "infinitesimal" phenomena. For example, let A be a local ring, $\mathfrak{M}$ its maximal ideal, X a scheme on Spec(A), and $X_0 = X \otimes_A A/\mathfrak{M}$ its "reduction modulo $\mathfrak{M}$." The schemes $X_n = X \otimes_A A/\mathfrak{M}^{n+1}$, which may be called the "infinitesimal neighborhoods" of X_0, can be considered as "intermediaries" between X_0 and X; X_n is a scheme over an *artinian* ring $A/\mathfrak{M}^{n+1}$ whose spectrum is reduced to a single point as the radical $\mathfrak{M}/\mathfrak{M}^{n+1}$ is nilpotent.

39. Pursuing this idea, it is natural to consider simultaneously *all* the infinitesimal neighborhoods X_n of X_0, which are schemes all having the same *underlying space* as X_0; the structure sheaves $\mathcal{O}_{X_n}$ form, in a natural way, a projective system, and, thus, a new structure comes into consideration, that of a space X_0 equipped with a projective system of sheaves of rings. This is a particular case of the notion of *formal scheme*, which in Grothendieck's theory corresponds to the idea of *completion* in topological algebra.

The starting point is a noetherian scheme X and a closed subset X′ of X, defined by a subsheaf of ideals $\mathcal{J}$ of $\mathcal{O}_X$; equipped with the sheaf of rings $\mathcal{O}_X/\mathcal{J}$, X′ becomes a closed subscheme X_0 of X. The scheme X_n, obtained by equipping the space X′ with the sheaf of rings $\mathcal{O}_X\mathcal{J}^{n+1}$, can also be considered, and X′ equipped with the projective limit of this projective system of sheaves of rings, is called the *formal completion* of X along X′, and is denoted by $X_{/X'}$. For a coherent sheaf $\mathcal{F}$ on X, the sheaves $\mathcal{F}_n = \mathcal{F} \otimes_{\mathcal{O}_X} (\mathcal{O}_X/\mathcal{J}^{n+1})$ on X′ are considered, and their projective limit is a coherent sheaf $\mathcal{F}_{/X'}$ on $X_{/X'}$, called the *completion* of $\mathcal{F}$. Its sections over X′ are called the *formal sections* of $\mathcal{F}$ over X′; for $\mathcal{F} = \mathcal{O}_X$, they are precisely the "holomorphic functions along X′" from Zariski's theory (when X is a variety over a field, cf. (VII, 46)).

40. Using these ideas and following a conjecture of Serre, Grothendieck was able to give a *cohomological* proof generalizing Zariski's "theorem of holomorphic functions" (VII, 46) to the theory of schemes (while simplifying it), and implying a very general form of Zariski's "connectedness theorem" and of Zariski's "main theorem" (VII, 43).

The starting point of this generalization is the consideration of two noetherian schemes X, Y and of a *proper* morphism $f: X \to Y$, a notion introduced by Chevalley, which is the "relativization" of the idea of complete variety in the sense of Weil (VII, 38). A closed subset Y′ of Y, its inverse image $X' = f^{-1}(Y')$, and the formal completions $\hat{X} = X_{/X'}$, $\hat{Y} = Y_{/Y'}$ are considered; from the morphism f is deduced a morphism $\hat{f}: \hat{X} \to \hat{Y}$ of formal schemes, which is evidently the "projective limit" of the morphisms $f_n: X_n \to Y_n$ of the "infinitesimal neighborhoods" of X′ and Y′ deduced from f by base change. That being so, let $\mathscr{F}$ be a coherent sheaf on X, $\mathscr{F}_{/X'}$ its completion along X′; then the $R^q\hat{f}_*(\mathscr{F}_{/X'})$, (VIII, 31) are coherent sheaves on $\hat{Y}$, the $R^q f_*(\mathscr{F})$ are coherent sheaves on Y, and there are canonical isomorphisms:

$$R^q\hat{f}_*(\mathscr{F}_{/X'}) \simeq \varprojlim_n R^q(f_n)_*(\mathscr{F}_n) \simeq (R^q f_*(\mathscr{F}))_{/Y'}.$$

The case $q = 0$ gives back Zariski's fundamental theorem on "holomorphic functions" and the connectedness theorem in the following general form (Stein factorization): f factors into $X \xrightarrow{h} Z \xrightarrow{g} Y$, where g is a finite morphism and the fibers of h are connected.

41. The case considered first in (VIII, 39), where, in addition, the local ring A is assumed complete, is the one where Y = Spec(A), Y′ being reduced to the unique closed point. Then the question arises of whether a scheme X_0 proper over the field $k = A/\mathfrak{M}$ is the fiber of a proper morphism X → Spec A at the closed point. Even by requiring in addition that this morphism be flat and by trying only to "lift" X_0 to a formal scheme $\hat{X}$, it can be verified that there are "obstructions" of a cohomological nature to this lifting. Grothendieck has proved that the lifting is possible if X_0 is a *curve* (i.e. of dimension 1), but Serre and Mumford have given examples where the lifting is not possible in dimension > 1, even when A is a ring of p-adic numbers; the possibility of lifting in this case is particularly interesting because the passage is from a base field of characteristic $p > 0$ to a field of characteristic 0 where the "transcendental" methods of algebraic geometry are applicable.

42. The theory of the varieties of Serre is conveniently contained in the theory of schemes: an affine Serre variety V over an algebraically closed field k corresponds to a reduced k-algebra of finite type, the points of V being identified with the maximal ideals of A; these are simply the *closed* points of the scheme Spec(A). They can still be identified with k-homomorphisms of algebras $A \to k$. Similarly, if it is desired to consider the "points of V with values in K," where K is an arbitrary extension of k (VII, 25), it is sufficient to identify them with the k-homomorphisms $A \to K$, K being considered as k-algebra. This idea has been considerably generalized by Grothendieck: given a S-scheme X, the *morphisms* T → X that are S-morphisms, that is, morphisms such that the composite morphism T → X → S is the structure morphism of T, are called "points of X with values in the S-scheme T" (or the T-points of X). We designate the set of these S-morphisms by $\mathrm{Mor}_S(T, X)$; Grothendieck's idea consists in associating to each S-scheme T, the set $\mathrm{Mor}_S(T, X)$, and in noting that a *functor* (contravariant) $\mathsf{h}_X : (\boldsymbol{Sch})_{/S} \to \boldsymbol{Ens}$ from the category of S-schemes to the category of sets is thus defined, because to each S-morphism $u: T' \to T$ there is a corresponding map

$\mathsf{h}_X(u): \mathrm{Mor}_S(T, X) \to \mathrm{Mor}_S(T', X)$ that associates to each S-morphism $T \to X$, the composite $T' \xrightarrow{u} T \to X$. The S-scheme X is said to *represent* the functor h_X, and it can be shown that the knowledge of the functor h_X *determines* the S-scheme X, up to isomorphism.

43. A remarkable fact is that it is often easier to define the functor h_X than the scheme X itself; whence comes the very original idea of Grothendieck that when it is a matter of constructing certain schemes satisfying given conditions, to define the corresponding functor instead of the scheme itself, on condition, of course, that it is known in some way or other that this functor is *representable* by an S-scheme.

For example, around 1950, the question was posed of defining a "projective space" $\mathbf{P}_n(A)$ when A is any ring (having divisors of zero in general), and diverse inadequate definitions were proposed. The suitable definition comes through the corresponding functor: to every affine A-scheme, Spec(B), where B is an A-algebra, is associated the set of the "hyperplanes" of B^{n+1}, defined as the kernels of the B-linear maps $B^{n+1} \to B$; this functor is representable by the desired A-scheme $\mathbf{P}_n(A)$. In fact, this definition generalizes, and for every scheme S and every quasi-coherent $\mathcal{O}_S$-Module $\mathcal{E}$, a "projective scheme" $\mathbf{P}(\mathcal{E})$ associated to $\mathcal{E}$ can be defined. This permits the definition of the notion of projective S-scheme as a closed subscheme of a $\mathbf{P}(\mathcal{E})$. The *grassmannians* $\mathbf{G}_m(\mathcal{E})$ ($\mathbf{P}(\mathcal{E})$ is $\mathbf{G}_1(\mathcal{E})$) are defined in an analogous way. Then Grothendieck avails himself of the grassmannians to resolve a more difficult problem: given an S-scheme X, an S-scheme is sought that "parametrizes" the closed subschemes Z of X that are flat over S. The corresponding functor must associate to every S-scheme T, the set of closed subschemes of $X_{(T)} = X \times_S T$ that are flat over T. It can be shown that if S is noetherian and X projective over S, then the preceding functor is representable by an S-scheme $\mathbf{Hilb}_{X/S}$, called the *Hilbert scheme* of X, a disjoint sum of projective S-schemes. These schemes play a role analogous to that of the "Chow coordinates" (VII, 31). Under stricter conditions, a *Picard scheme* $\mathbf{Pic}_{X/S}$, which "parametrizes" the classes of "divisors" on X, can be defined as well.

44. The same idea of passing to the functor h_X serves to define additional structures on a scheme X. For example, to define the notion of "equivalence relation" on a scheme X, on each set $\mathrm{Mor}_S(T, X)$, an equivalence relation in the usual sense is considered that must "depend functorially" on T. Similarly, to define X as a "group scheme" or "ring scheme," etc., the structure of group, of ring, etc., "depending functorially" on T is defined on each set $\mathrm{Mor}_S(T, X)$.

45. The most profound of the new ideas from the theory of schemes derives also from base change. In the theory of algebraic varieties over an algebraically closed field k, Weil had defined in 1949, not only the notion of vector bundle, but also that of *principal bundle* over such a variety V, by copying the usual definition from differential geometry, the group G operating on such a bundle P being assumed algebraic (and operating so that the map $G \times P \to P$ is a morphism), and the

bundle assumed "locally trivial" for the Zariski topology. In 1958, Serre remarked that this definition does not have the anticipated properties; for example, in the theory of Lie groups, if H is a closed subgroup of a Lie group G, then G is naturally equipped with the structure of principal bundle (of group H) over the variety G/H. In the theory of algebraic groups over k, for an algebraic group G and a closed subgroup H, G/H can still be defined as an algebraic variety, but for the natural operation of H over G, G is no longer, in general, a principal bundle over G/H in the sense of Weil's definition.

However, Serre observed that, in this case, the local "triviality" is recovered if the Zariski open sets in the base, which enter in the definition of local triviality, are replaced by suitable finite étale coverings (VIII, 34) of these open sets, the bundle being replaced by its "inverse image" above this covering.

46. Starting from this remark, Grothendieck conceived the idea of replacing the Zariski topology on a scheme X by a new structure, called the *étale topology*, which is no longer a topology in the usual sense at all: here the usual open sets U of S (or rather their canonical injections U → S) are replaced by *finite étale morphisms* V → U (U being an open set of X); it can be said that the open sets are now "outside the space" instead of being subsets of it. The interest of this definition is that the notions of *sheaf* and of *cohomology with values in a sheaf* can be transferred to this "topology," and this "étale cohomology" has properties that bring it much closer to the topological cohomology of the classical case than the sheaf cohomology for the Zariski topology when, for example, the scheme X is a variety over a field of characteristic $p > 0$ (see (IX, 85)).

47. The construction of schemes "representing" functors (VIII, 42) is not always possible; for example, Nagata has constructed a complete, nonprojective variety X over a field k, for which the Hilbert functor is not representable. To palliate these inconveniences, M. Artin has introduced objects more general than schemes, which he calls *algebraic spaces*, and that can be conceived as "quotient spaces" of schemes (by equivalence relations not necessarily giving a "quotient scheme"). Most properties of schemes extend to these spaces and, in addition, most constructions (quotients, Hilbert scheme, Picard scheme, etc.) are made in this category without awkward restriction.

IX — RECENT RESULTS AND OPEN PROBLEMS

1. Since 1960, there has been a tremendous increase in the number of mathematicians who are working in algebraic geometry. A summer institute held in California in 1974 was attended by 270 mathematicians! This phenomenon can be attributed on one hand to the unusual wealth and diversity of (old and new) interesting and "natural" problems afforded by algebraic geometry, and to their multiple connections with number theory, analytic geometry, and analysis, and on the other hand to the availability of powerful and easy-to-grasp techniques of attack, with no need for an appeal to some personal and more or less reliable "intuition."

It is of course utterly impossible to give an exhaustive treatment of all the results recently obtained in algebraic geometry. Here we can give only sketchy accounts of what seem to be the more significant developments, supplemented by bibliographical references to the original papers, or more often to expository books and papers, which themselves usually contain a much more extensive bibliography.

The division of this chapter into sections is purely for convenience, and their multiple interconnections will easily be perceived by the reader.

1. Problems on curves

2. Many papers on algebraic geometry published since 1950 are concerned with problems on curves and surfaces in projective complex spaces $\mathbf{P}_n(\mathbf{C})$ that already had been studied by the Italian geometers. But many of the Italian proofs were inconclusive, or based on "intuitive" arguments, and it was necessary to put them on more secure foundations, chiefly based on sheaf cohomology and other modern techniques, such as the theory of abelian varieties. Another problem was to see if the results of the Italians generalized to projective curves or surfaces over an arbitrary algebraically closed field of characteristic $p > 0$.

3. *A) Special Divisors.* Let C be a smooth projective complex curve of genus $g \geqslant 2$. For any divisor D on C, let $r(\mathrm{D}) = l(\mathrm{D}) - 1$ be the dimension of the projective space $|\mathrm{D}|$ (VI, 9). Recall that if Δ is a canonical divisor on C (VI, 13), the Riemann-Roch theorem (VI, 14) gives

$$r(\mathrm{D}) = d - g + l(\Delta - \mathrm{D}) \tag{1}$$

where $d = \deg D$. The divisor D is called *special* if $l(\Delta - D) > 0$. For $d > 0$, Clifford's theorem (VI, 14) shows that

$$r(D) \leqslant d/2 \leqslant g - 1. \tag{2}$$

One may show that on a hyperelliptic curve (VI, 15), for all pairs (r, d) such that $r \leqslant d/2 \leqslant g - 1$, there exist divisors D of degree d with $r(D) = r$. But what can be said for special divisors on a nonhyperelliptic curve C ($g \geqslant 3$); do they exist and "how many" of them are there? This problem occupied several mathematicians in the period 1960–1980, and it seems worthwhile to describe its solution in some detail, in order to exhibit the constant interplay between the geometric insights of the Italians and the modern tools of the theory of schemes that made them effective.

4. To put these questions in a precise form, we introduce the *symmetric product* C_d of d copies of C (the variety of orbits of the symmetric group $\mathfrak{S}_d$ acting naturally on the product C^d), which parametrizes the set of *all* divisors of degree d on C. Let C_d^r be the subvariety of C_d consisting of the divisors D such that $r(D) \geqslant r$, and let W_d^r be the natural image of C_d^r in the jacobian J(C) (variety of classes of divisors of C_d^r for linear equivalence (VII, 57)). The precise form of the problem of special divisors is then to determine the *dimensions* of C_d^r and W_d^r for $r \leqslant d/2 \leqslant g - 1$.

5. Using a different language, the problem had been tackled by Brill and Noether in 1874. If

$$\rho = g - (r + 1)(g - d + r) \tag{3}$$

they proved that, if $\rho \geqslant 0$, then, *assuming* $W_d^r \neq \emptyset$,

$$\dim W_d^r \geqslant \rho \qquad \text{and} \qquad \dim C_d^r \geqslant \rho + r. \tag{4}$$

Their proof can be translated with unessential modifications into the language of sheaf cohomology, and over an algebraically closed field of arbitrary characteristic. But until 1970, the fact that $W_d^r \neq \emptyset$ for $\rho \geqslant 0$ had been proved only for particular values of r and d. The general proof was obtained in 1971 by Kempf, Kleiman, and Laksov, using all the heavy machinery of the theory of schemes (Hilbert schemes, Chow rings, Chern classes, Thom polynomials). With this technique, they were able to compute the cohomology class w_d^r of W_d^r in the cohomology ring of J(C), and to prove that this class is $\neq 0$ when $\rho \geqslant 0$; hence $W_d^r \neq \emptyset$.

6. Examples abound of curves for which $\dim W_d^r > \rho$. Brill and Noether had assumed without proof that for a curve with "general moduli" (VI, 27 and IX, 14) $\dim W_d^r = \rho$. Petri, in (Über Spezialkurven I. Math. Ann. 93, 182–209 (1924)), even claimed in effect that it was "well-known" that W_d^r was in this case smooth off W_d^{r+1}, or, what comes to the same thing, that the variety G_d^r of all linear series of dimension r and degree d, is smooth.

Investigating the statements about W_d^r in 1976, Kleiman showed that this would follow from a geometric result on systems of chords of a rational normal curve (VI, 29), for which Severi had tried to give a proof based on a previous idea of Castelnuovo, but which turned out to be inconclusive. These ideas rely on an interesting geometric interpretation of the Riemann-Roch theorem. Let $C \subset \mathbf{P}_{g-1}$ be a canonical curve of genus g (VI, 26), and let $D = \sum_{i=1}^{d} P_i$ be a divisor of degree d on C; then $r(D) \geqslant r$ if and only if the points P_i belong to a $(d-r-1)$-dimensional *linear variety* of $\mathbf{P}_{g-1}$.

7. In 1889, Castelnuovo was considering the case $\rho = 0$, and wanted to compute the number (that he assumed to be finite) of elements of W^r_d for a smooth curve of genus g with "general moduli." He had the idea of considering a family (C_t) of curves depending on a parameter t, such that for $t \neq 0$, C_t is a canonical curve of genus g in $\mathbf{P}_{g-1}$, but for $t = 0$ there is a "degeneracy," C_0 being a *rational* curve with g nodes. Castlenuovo thought that the number of elements of W^r_d would be the same for all values of t, and that he could therefore reduce the problem of its computation to the case of the "Castelnuovo curve" C_0.

In 1921, Severi considered the general case $\rho \geqslant 0$, and in order to prove that $\dim W^r_d \leqslant \rho$ for a curve with "general moduli," he tried to reduce the proof to the "Castelnuovo curve" C_0. His arguments used a geometric interpretation of divisors on C_0 (also due to Castelnuovo): one considers the normalization (VII, 44) $\tilde{C}_0$ of C_0, which may be identified with the "normal" rational curve of degree d in $\mathbf{P}_d$ (VI, 29). For a divisor $D \in C^r_d$ on C_0, the linear series $|D|$ is lifted back to $\tilde{C}_0$ as the series cut out by hyperplanes of $\mathbf{P}_d$ containing a fixed linear variety Λ of dimension $d - r - 1$; in addition, Λ must meet each one of the g chords joining the pair of points (p_α, q_α) $(1 \leqslant \alpha \leqslant g)$ of $\tilde{C}_0$ that project on the g nodes of C_0.

8. It is this interpretation that Kleiman used in 1976 (again with the help of heavy machinery from the theory of schemes) to arrive at the following conclusion: the inequality $\dim W^r_d \leqslant \rho$ holds for curves with "general moduli" if what was called the "Castelnuovo-Severi-Kleiman" conjecture (abbreviated CSK) is true, namely:

(CSK) Let Γ be a normal rational curve of degree $d > 1$ in $\mathbf{P}_d$, and let $G(k, d)$ be the grassmannian of linear varieties of dimension $k < d$ in $\mathbf{P}_d$. Consider g generic chords of Γ; the subvariety of $G(k, d)$ consisting of linear varieties that meet each of these g chords has dimension at most $(k + 1)(d - k) - g(d - k - 1)$.

Finally in 1979, Griffiths and Harris gave a more geometric proof of Kleiman's result, and then proved CSK by using the Schubert calculus (IX, 76), thus solving the Brill-Noether problem. More precisely, their idea was to compare CSK for generic chords of Γ to "degenerate" systems of such chords. The statement of Petri about G^r_d for a curve with general moduli was clarified and proved in a special case by Arbarello and Cornalba, and then proved in generality by Gieseker (for the simplest proof, and references, see "A simpler proof of the Gieseker-Petri Theorem," D. Eisenbud and J. Harris, Inv. Math 74(1983) 269–280). The central technical advance here was the use of degeneration to certain *reducible* curves, as used in the study of moduli described below.

9. *B) Degree and Genus of Curves in* $\mathbf{P}_3$. We have seen (VI, 28) that in 1889, Castelnuovo had given inequalities for the genus of a smooth irreducible curve of given degree contained in a projective space $\mathbf{P}_r$. But even earlier, the problem had arisen of the *exact* determination of the pairs (d, g) formed by the degree d and the genus g of a smooth irreducible curve C in *ordinary* projective space $\mathbf{P}_3$. In 1882, Halphen published a long paper on this subject. It was then known that if C is contained in a quadric and not in a plane, there are two integers $a > 0, b > 0$ such that $d = a + b$ and $g = (a-1)(b-1)$; and conversely, for any pair (a, b) of integers > 0, there exists a smooth irreducible curve on a quadric, with the value $(a+b, (a-1)(b-1))$ of the pair (d, g). Halphen showed that if C is not contained in a plane nor in a quadric, then

$$0 \leqslant g \leqslant \frac{1}{6}d(d-3)+1 \tag{5}$$

and that for any pair (d, g) with $d > 0$ and g satisfying (5), there exists a smooth irreducible curve of degree d and genus g in $\mathbf{P}_3$; he even thought that there always existed such a curve on a cubic surface.

10. However, examples are now known of curves of degree 10 and genus 1 that are not on any (smooth or not) cubic surface; even Halphen's proof of (5) is not correct. The problem has recently been completely solved by Gruson and Peskine; over an algebraically closed field of characteristic 0, they have given a correct proof of (5), and established the following two existence theorems:

Theorem 1. For any pair (d, g) such that $d > 0$ and

$$\frac{1}{\sqrt{3}}d^{3/2} - d + 1 < g \leqslant \frac{1}{6}d(d-3)+1 \tag{6}$$

there exists a smooth irreducible curve of degree d and genus g on a smooth cubic surface.

Theorem 2. For any pair (d, g) such that $d > 0$ and

$$0 \leqslant g \leqslant \frac{1}{8}(d-1)^2 \tag{7}$$

there exists a smooth irreducible curve of degree d and genus g on a rational quartic surface having a double line.

11. The proofs rest on the generation of rational surfaces by blowing up points of $\mathbf{P}_2$ (VI, 51 and 52). For a cubic surface X, this gives an explicit $\mathbf{Z}$-basis of 7 elements for Pic(X) (VIII, 10), so that each class of divisors is represented by a system $(a, b_1, \ldots, b_6)$ of 7 integers. Moreover, if $b_1 \geqslant b_2 \geqslant \cdots \geqslant b_6 \geqslant 0$ and $a \geqslant b_1 + b_2 + b_3$, the class contains a smooth irreducible curve for which

(8) $$d = 3a - \sum_{i=1}^{6} b_i \quad \text{and} \quad g = \frac{1}{2}(a^2 - \sum_{i=1}^{6} b_i^2 - d) + 1$$

and the problem is reduced to a question of pure number theory, which is solved by using the theory of integral quadratic forms in 5 variables. The proof of Theorem 2 is similar but more complicated, involving the surface S obtained by blowing up 9 points of $\mathbf{P}_2$, and then its image X in $\mathbf{P}_3$ by a finite morphism. The proof requires the study of linear systems of curves on S, and the arithmetical properties of some special types of integral quadratic forms in 10 variables.

Finally, Hartshorne has shown how the methods of Gruson and Peskine may be generalized to treat the case of curves over an algebraically closed field of characteristic > 0. The analogous questions for curves in $\mathbf{P}^r (r > 3)$ are still largely open; see the book "Curves in Projective Space" by Joe Harris, Sem. de Math. Sup., Les Presses de l'Université de Montréal, 1982.

12. *C) "Moduli" of Curves.* The theory of "moduli" of algebraic curves, started by Riemann (VI, 27), had been deeply studied by the Italian geometers. Enriques and Severi considered the set Σ of plane irreducible curves of degree n having as only singularities d nodes, and Severi had stated that Σ was an irreducible family of cycles (VII, 32), hence a projective algebraic variety of dimension $3n + g - 1$ (g being the genus), "fibered" by the subvarieties consisting of the curves of Σ birationally equivalent to one of them. Since these subvarieties are of dimension $3n - 2g + 2$ if $g \geqslant 2$, this gave the dimension $3g - 3$ for the "quotient" variety. But Severi's proof of the irreducibility of Σ was not correct.

13. Around 1940, Teichmüller attacked the problem from the point of view of analytic geometry. One has to "parametrize" the isomorphism classes of Riemann surfaces of given genus g. On each such surface S, Teichmüller considers the classes of homotopically equivalent continuous maps of S into a fixed oriented compact surface T_0 of genus g, and among these classes, those containing at least one diffeomorphism preserving orientation. For $g \geqslant 2$, he showed that the set Θ of isomorphism classes of pairs (S, ξ) for a point $\xi \in S$ (an isomorphism $(S, \xi) \tilde{\to} (S', \xi')$ being an isomorphism of holomorphic manifolds sending ξ to ξ') is naturally equipped with a topology for which it is homeomorphic to an open ball in $\mathbf{R}^{6g-6}$. Later, using the Kodaire-Spencer theory of "deformations of complex structures," L. Bers could show that the "Teichmüller space" Θ may be given the structure of a complex manifold of complex dimension $3g - 3$. In that manifold operates a discrete group γ consisting of equivalence classes of isomorphisms of Riemann surfaces: two isomorphisms are in the same class if they leave invariant a given point of Θ. One may show that γ operates holomorphically and properly on Θ, and the "orbit space" Θ/γ is an analytic space of dimension $3g - 3$, which deserves to be called the "space of moduli" of Riemann surfaces of genus g; but in general it has singular points (in particular the points that are isomorphism classes of hyperelliptic curves (VI, 15)).

14. In 1965, Mumford succeeded in constructing a "scheme of moduli" for smooth curves of genus g, defined over *any* algebraically closed field. Such a scheme M_g should have the property that, for any base change $\mathbf{Z} \to k$ (VIII, 32), where k is any algebraically closed field, the closed points of the scheme $M_g \otimes_{\mathbf{Z}} k$ be functorially identified with the isomorphism classes of smooth irreducible curves of genus g over k. The basic idea, as in Teichmüller's theory, is to start with a richer structure on curves defined over k, namely the "tricanonical" embedding (VI, 26) in a projective space $\mathbf{P}_n(k)$, with $n = 5g - 6$; two such embeddings correspond to birationally equivalent curves if and only if they are deduced from one another by a projective transformation. Mumford shows that for these structures it is possible to define a scheme H_g that plays the role of M_g and is a closed subscheme of the Hilbert scheme of $\mathbf{P}_n(\mathbf{Z})$ (VIII, 43). The group scheme $PGL(n)$ of automorphisms of $\mathbf{P}_n(\mathbf{Z})$ operates naturally on H_g and the problem is to obtain a "quotient scheme" (VIII, 44), which would be the scheme of moduli M_g. By a skillful use of invariant theory, Mumford first obtains a quotient of the scheme $H_g \otimes_{\mathbf{Z}} \mathbf{Q}$ over $\text{Spec}(\mathbf{Q})$ (which corresponds to $M_g \otimes_{\mathbf{Z}} \mathbf{Q}$). To get M_g itself, he uses as intermediate scheme a "scheme of moduli" A for *abelian varieties* (VII, 60) equipped with additional structures ("rigidifications" and "polarizations") (IX, 113 and 115). The final step is based on a theorem of Torelli that characterizes a curve up to isomorphism by its "polarized" jacobian variety (IX, 116). Again using invariant theory, Mumford finally proves that the set of orbits in H_g under the action of $PGL(n)$ can be identified with a subscheme of the "scheme of moduli" A, proving thus the existence of M_g.

More recently, Deligne and Mumford have been able to revive partially Severi's idea and to prove that $M_g \otimes_{\mathbf{Z}} k$ is *irreducible* for any algebraically closed field k.

15. When $g = 1$, the unique "modulus" of an elliptic curve (over $\mathbf{C}$) is identified with the set of ratios of two generators of the module of periods of an elliptic function corresponding to the curve (VII, 57). Equivalently, this set is the orbit of a complex number τ such that $\text{IM}\,\tau > 0$ under the action of the group $SL(2, \mathbf{Z})$ on the positive half-plane H of $\mathbf{C}$: $z \mapsto (az + b)/(cz + d)$ with a, b, c, d integers such that $ad - bc = 1$. The "variety of moduli" is thus identified with the space $H/SL(2, \mathbf{Z})$ of these orbits. This fact can be generalized to families of abelian varieties equipped with additional structures, which are naturally "parametrized" by spaces of orbits S/Γ, where S is a bounded symmetric domain of a $\mathbf{C}^n$ and Γ a discrete group acting on S. This conception links the theory of abelian varieties to the theory of automorphic functions, and is the source of numerous problems in number theory, generalizing the complex multiplication of elliptic functions (VII, 57).

In particular, it follows that the moduli space of elliptic curves is the affine line $\mathbf{A}^1$. It was proved by Severi that for $g \leqslant 10$, M_g is at least the image of a rational variety, and he conjectured that this would be true for all g. Recently Sernesi and Ran-Chang have shown that this does hold for $g \leqslant 13$, but Eisenbud, Harris, and Mumford have shown that it fails, quite badly for all $g \geqslant 23$. The intermediate cases are still open. (See for example "Systèmes Pluricanoniques sur l'Espace des modules des courbes et diviseurs de courbes k-gonales, Maurizio Cornalba, Sem. Bourbaki no. 615, November 1983, for a survey.)

2. Problems on surfaces

16. We have already seen in Chapters VII and VIII that it is almost impossible to dissociate the theory of smooth algebraic varieties over **C** from analytic geometry, in particular from the theory of compact complex manifolds. We shall see that this fruitful symbiosis has been continued in more recent work.

17. *A) Exceptional Curves and Minimal Models.* The process of *blowing up* a point of $\mathbf{P}_n(\mathbf{C})$ (VI, 50) can be extended to all smooth complex manifolds. Blowing up a point x_0 of such an n-dimensional manifold X yields another complex n-dimensional manifold Y and an analytic map $f: \mathrm{Y} \to \mathrm{X}$, such that the restriction of f to $\mathrm{Y} - f^{-1}(x_0)$ is an *isomorphism* onto $\mathrm{X} - \{x_0\}$, and $\mathrm{Z} = f^{-1}(x_0)$ is an $(n-1)$-dimensional manifold; f is called the *blowing down* of Z to a point x_0.

If X is a surface, then $f^{-1}(x_0) = \mathrm{C}$ is a smooth curve, and $(\mathrm{C}\,.\,\mathrm{C}) = -1$ (VI, 37). If in addition X is algebraic, then so is Y; it is birationally equivalent to X, and C is a *rational* curve. A famous result of Castelnuovo is that if Y is a smooth algebraic surface, for a smooth curve C on Y to be blown down by a suitable morphism, the conditions that C is rational and $(\mathrm{C}\,.\,\mathrm{C}) = -1$ are also *sufficient*; later, the Castelnuovo criterion was extended to surfaces defined over fields of characteristic > 0.

18. The Italians called the curves satisfying Castelnuovo's criterion *exceptional curves of the first kind*; a smooth surface is called *minimal* if it contains no such curves. Zariski has proved that for any smooth algebraic surface X, it is possible to blow down all exceptional curves of the first kind (which are finite in number) to obtain a minimal surface Y birationally equivalent to X. Furthermore, if X is not a ruled surface (IX, 22), the minimal model Y is *unique* up to isomorphism.

19. *B) The Enriques Classification.* We have seen (VI, 49) how, after Enriques had introduced the plurigenera of a surface, he and Castelnuovo obtained striking results, showing how the knowledge of these numbers and of the irregularity $q = p_g - p_a$ could characterize whole classes of surfaces, such as rational or ruled ones. After 1906, Enriques continued that study, and in 1914 arrived at a division of all surfaces into four classes, characterized by their plurigenera:

- Class I) $\mathrm{P}_{12} = 0$, or equivalently $\mathrm{P}_n = 0$ for all $n \geqslant 1$.
- Class II) $\mathrm{P}_{12} = 1$, or equivalently $\mathrm{P}_n = 0$ or 1 for all $n \geqslant 1$.
- Class III) $\mathrm{P}_{12} \geqslant 2$ and $\mathrm{P}_n \sim \mathrm{C}\,.\,n$ for a constant $\mathrm{C} \neq 0$ when n tends to $+\infty$.
- Class IV) $\mathrm{P}_{12} \geqslant 2$ and $\mathrm{P}_n \sim \mathrm{C}\,.\,n^2$ for a constant $\mathrm{C} \neq 0$ when n tends to $+\infty$.

In addition he was able to obtain geometric descriptions of many subclasses of these classes.

Around 1960 there was a renewed interest in these questions. Zariski and Shafarevich and his school put the arguments of Castelnuovo and Enriques on more secure foundations than their "intuitive" proofs. Zariski, M. Artin, Mumford, and Bombieri extended the Enriques classification to surfaces over a field of characteristic $p > 0$. Finally, Kodaira, using "transcendental" methods in a remarkable series of papers, enlarged the classification problem to *all compact complex surfaces*, and in so doing discovered many new properties of surfaces of classes II and III.

20. We recall that for a smooth projective algebraic irreducible surface X, q and the P_n are absolute birational invariants (VI, 32), which are now defined as dimensions of cohomology groups of sheaves over the base field (algebraically closed) (VIII, 11–16 and 21). Other invariants linked to these are the Betti numbers $R_j (0 \leqslant j \leqslant 4)$ of the surface. One has $R_0 = R_4 = 1$, $R_1 = R_3 = 2q$, but R_2 is not an absolute birational invariant; it is linked to the self-intersection number $(\Delta . \Delta)$ of a canonical divisor Δ, and to the Euler-Poincaré characteristic $\chi(\mathcal{O}_X)$ (VIII, 12) by M. Noether's formula

$$\chi(\mathcal{O}_X) = \frac{1}{12}(e(X) + (\Delta . \Delta)) \tag{9}$$

where $e(X) = R_0 - R_1 + R_2 - R_3 + R_4$ is the topological Euler-Poincaré characteristic. Finally, the two Chern classes (VIII, 5) c_1 and c_2 are given by

$$\langle c_1^2, X \rangle = (\Delta . \Delta) \quad \text{and} \quad \langle c_2, X \rangle = e(X). \tag{10}$$

It is useful to remember that when Y is a surface obtained by blowing up a point of X, then

$$R_2(Y) = R_2(X) + 1, \qquad e(Y) = e(X) + 1, \qquad (\Delta_Y . \Delta_Y) = (\Delta_X . \Delta_X) - 1. \tag{11}$$

21. Any linear system $|D|$ of positive divisors on X defines a rational map $\varphi_{|D|}$ of X into a projective space. To each point $x \in X$ not among the fixed points of $|D|$ (i.e. those that are in the support (VI, 43) of every divisor of the system), is associated the hyperplane in the projective space $|D|$ that consists of all divisors of $|D|$ whose support contains x. This hyperplane $\varphi_{|D|}(x)$ is an element of the projective space *dual* to $|D|$.

It can be shown that the Enriques classification may also be expressed by considering the maps $\varphi_{|n\Delta|}$ for multiples of the canonical divisor Δ. One defines the *Kodaira dimension* $\kappa(X)$ or Kod(X) of X, as follows: if $|n\Delta| = \emptyset$ for all integers $n \geqslant 1$, one takes $\kappa(X) = -\infty$ by convention; otherwise, $\kappa(X)$ is the *maximum dimension* of the images $\varphi_{|n\Delta|}(X)$ for $n \geqslant 1$. Then the four classes of the Enriques classification correspond to the values $\kappa(X) = -\infty, 0, 1, 2$, respectively. We summarize the main results known on the smooth irreducible surfaces of these four classes (most of the results having first been formulated by Castelnuovo and Enriques).

22. Class I) $\kappa(X) = -\infty$. In this case, X is *rational* (i.e. birationally equivalent to $\mathbf{P}_2$) or *ruled* (i.e. birationally equivalent to a product $\mathbf{P}_1 \times C$ where C is a smooth irreducible curve). A very large number of these surfaces, distinguished by various geometrical properties, have been studied during the nineteenth century. The *minimal* rational surfaces (IX, 18) are $\mathbf{P}_2$ and the surfaces $\mathbf{P}(\mathscr{E})$ (VIII, 43), where $\mathscr{E}$ is an $\mathcal{O}_{\mathbf{P}_1}$-Module $\mathcal{O}_{\mathbf{P}_1} \oplus \mathcal{O}_{\mathbf{P}_1}(n)$ (VIII, 21) for an integer $n \geqslant 1$.

The minimal ruled surfaces are the $\mathbf{P}(\mathscr{E})$, where $\mathscr{E}$ is a locally free (IX, 35) $\mathcal{O}_C$-Module of rank 2 on a curve C of genus $g \geqslant 1$. For these minimal ruled surfaces, one has $q = g$, $R_2 = 2$ and $(\Delta . \Delta) = 8(1-g)$.

23. Class II) $\kappa(X) = 0$. We only consider the *minimal* surfaces of this class; for these $(\Delta . \Delta) = 0$, hence $e(X) = 12\,\chi(\mathcal{O}_X)$. They are divided into four subclasses, according to the values of p_g and q.

Class IIa) $p_g = 0$, $q = 0$. These are called *Enriques surfaces*; the original (non-smooth) surface of degree 6 defined by Enriques (VI, 49) has a smooth model in this subclass. They satisfy $R_1 = 0$ and $R_2 = 10$.

Class IIb) $p_g = 0$, $q = 1$, hence $R_1 = 2$, $R_2 = 2$. These surfaces are called "hyperelliptic," or better "bielliptic." They are described as varieties of orbits $(E \times F)/G$, where E and F are two smooth elliptic curves (VII, 57) and G a finite commutative group acting on E by translations of the group E, and on F in such a way that F/G is a rational curve. The list of all possible such groups G was determined by Bagnera and de Franchis in 1907; there are 7 possibilities.

Class IIc) $p_g = 1$, $q = 0$, hence $R_1 = 0$, $R_2 = 22$. These surfaces are known as *K3 surfaces* (for Kummer, Kähler, and Kodaira, according to some texts). The first example of such a surface goes back to Kummer, and is a special case of surfaces obtained by the following procedure. One starts from an abelian surface A, which can be written $A = \mathbf{C}^2/L$, where L is a lattice isomorphic to $\mathbf{Z}^4$ (VII, 55), and then identifies in that surface each point t with its inverse $-t$. This gives a surface X_0 with 16 isolated double points, and blowing up these points yields a K3 surface. When A is the jacobian of a curve of genus 2, X_0 is the surface (of degree 4 in $\mathbf{P}_3$) originally studied by Kummer.

The K3 surfaces have been the subject of many papers, aimed in particular at the study of curves on these surfaces. The quotient of a K3 surface by an involution without fixed points is an Enriques surface, and all Enriques surfaces can be obtained in this way.

Class IId) $p_g = 1$, $q = 2$. These are the abelian surfaces (topologically tori $\mathbf{T}^4$).

24. Class III) $\kappa(X) = 1$. If X in this class is minimal, there exists a smooth irreducible curve B and a surjective morphism $\pi : X \to B$ such that the generic fiber $\pi^{-1}(\eta)$ (where η is the generic point (VIII, 30) of B) is an elliptic curve. Surfaces having this property (for any smooth irreducible curve B) are called *elliptic*. All elliptic surfaces are such that $\kappa(X) \leqslant 1$, but there are elliptic surfaces

with $\kappa(X) = -\infty$ or $\kappa(X) = 0$; for instance, all Enriques surfaces are elliptic. An example of a surface with $\kappa(X) = 1$ is given by the product $C \times D$ of two smooth curves, where C has a genus $\geqslant 2$ and D is elliptic.

25. Class IV) $\kappa(X) = 2$. The surfaces of this class are called *general*; although many examples are known (for instance the product $C \times D$ of two curves of genus $\geqslant 2$), much less is known about their structure than for the "special" surfaces of the three other classes. In contrast with the existence of the tricanonical mapping for curves (VI, 26), it is not always true that there exists an integer $n > 0$ such that the image of a general surface X by $\varphi_{|n\Delta|}$ is a smooth surface; what is true is that for $n \geqslant 5$, $\varphi_{|n\Delta|}$ is everywhere defined and $\varphi_{|n\Delta|}$ has only a finite number of isolated singular points.

A proposed division of general minimal surfaces X into subclasses is according to the values of the pairs $(\langle c_1^2, X\rangle, \langle c_2, X\rangle)$, but it is not yet known what are the pairs of integers (m, n) that may occur. It is known that one must have $n > 0$, that $m + n \equiv 0 \pmod{12}$, $n \geqslant 5m + 36$, and that there are infinitely many nonisomorphic surfaces such that $5m - n + 36 = 0$. Recently it has been proved that $\langle c_1^2, X\rangle \leqslant 3\langle c_2, X\rangle$ for minimal surfaces of general type, the constant 3 being best possible (equality holds for infinitely many pairs (m, n)).

An interesting subclass of general surfaces are those for which $\langle c_1^2, X\rangle = 1$, $\langle c_2, X\rangle = 11$; they all satisfy $p_g = q = 0$. The first example was found by Godeaux: take in $\mathbf{P}_3(\mathbf{C})$ the surface S′ of equation $x_1^5 + x_2^5 + x_3^5 + x_4^5 = 0$. The group $G = \mathbf{Z}/5\mathbf{Z}$ acts without fixed points on S′ by $(x_1, x_2, x_3, x_4) \mapsto (x_1, \varepsilon x_2, \varepsilon^2 x_3, \varepsilon^3 x_4)$ with $\varepsilon = e^{2\pi i/5}$ and the powers of that mapping, and $S = S'/G$ is the Godeaux surface.

26. The classification problem for surfaces over a field of characteristic $p > 0$ was examined by Bombieri and Mumford; it turns out that it does not substantially differ from the Enriques classification except for $p = 2$ or 3.

3. Varieties of arbitrary dimension

27. *A) Classification Problems.* Under this heading, we shall consider only smooth projective (or quasi-projective (IX, 34)) complex irreducible varieties. Since 1970, substantial progress has been achieved in the problem of classifying these varieties of dimension $\geqslant 3$ by trying to obtain results on the model of those known for dimensions 1 and 2. The difficulties are much greater, the first being non-existence of minimal models (IX, 18), so that all one can expect is a classification up to *birational equivalence*.

28. The main tools in this theory are still the maps $\varphi_{|n\Delta|}$ and the concept of Kodaira dimension, whose definitions are the same as for surfaces (IX, 21). A central notion is the concept of *algebraic fiber space* (which in general is *not* a fiber bundle in the topological sense). A morphism $f\colon V \to W$ of algebraic varieties defines V as a fiber space over W (one also says f itself *is* a fiber space) if f is

surjective and the fibers $f^{-1}(w)$ for $w \in W$ are all connected; the generic fiber F is then smooth and irreducible. An important result is the *Iitaka inequality*

$$\kappa(V) \leqslant \dim W + \kappa(F). \tag{12}$$

One again divides algebraic varieties V of dimension n into four classes:

Class I) $\kappa(V) = -\infty$, called varieties of *elliptic type.*
Class II) $\kappa(V) = 0$, called varieties of *parabolic type.*
Class III) Varieties with $\dim V > \kappa(V) \geqslant 1$.
Class IV) Varieties with $\kappa(V) = \dim V$, called varieties of *hyperbolic type.*

29. In contrast with the case of surfaces, almost no general result is known for varieties of elliptic type ($\kappa(V) = -\infty$). When $f: V \to W$ is a fiber space and $\kappa(F) = -\infty$ for the generic fiber F, Iitaka's inequality (12) shows that $\kappa(V) = -\infty$; this is the case in particular for ruled varieties (birationally equivalent to $\mathbf{P}_1 \times W$).

Another type of elliptic varieties consists of the *unirational* varieties of dimension $n \geqslant 3$; this means that for such a variety V, the field $\mathbf{C}(V)$ of rational functions on V (VII, 26) is a subfield of a purely transcendental extension $\mathbf{C}(T_1, T_2, \ldots, T_n)$; the latter can then be taken to be a *finite* extension of $\mathbf{C}(V)$. For $n = 1$ and $n = 2$, these varieties are in fact *rational*; this was proved by Lüroth for curves (VI, 42), and for surfaces it follows from Castelnuovo's criterion (VI, 49). For a long time it was not known if this result also was true for $n \geqslant 3$; the difficulty is that most birational invariants are the same for rational and unirational varieties. Finally in 1962, M. Artin and Mumford 1) showed that the torsion subgroup of the cohomology group $H^3(V; \mathbf{Z})$ is a birational invariant, and 2) constructed a unirational variety V of dimension 3 such that $H^3(V; \mathbf{Z}) \simeq \mathbf{Z}/2\mathbf{Z}$, showing that V is not rational. Simultaneously, two other examples were found: Clemens and Griffiths showed that in $\mathbf{P}_4$ a cubic hypersurface V is unirational but not rational by considering its intermediate jacobian $J_1(V)$ (VII, 58), and Manin and Iskovskih gave an example of quartic unirational hypersurface V in $\mathbf{P}_4$ that is not rational, by studying the group of $\mathbf{C}$-automorphisms of $\mathbf{C}(V)$.

30. An important tool in classification theory is the *Albanese map* $\alpha: V \to A(V)$ (VII, 59). For varieties V of parabolic type ($\kappa(V) = 0$), this map is a *fiber space*: as one has $q(V) = \dim A(V)$ for the irregularity $q(V)$ (VII, 59), one has $q(V) \leqslant \dim V$. The equality $q(V) = \dim V$ (for $\kappa(V) = 0$) holds if and only if V is birationally equivalent to an abelian variety. If $\kappa(V) = 0$ and $q(V) = \dim V - 1$, the fiber F is such that $\kappa(F) = 0$.

31. For varieties V with $\kappa(V) \geqslant 1$ (classes III and IV), there is a general theorem of Iitaka: there exists a smooth variety V′ birationally equivalent to V and a fiber space $f: V' \to W$ such that $\dim W = \kappa(V)$ and $\kappa(F) = 0$ for the generic fiber F, f being unique up to birational equivalence.

32. A central conjecture in classification theory has been formulated by Ueno and Iitaka:

(C_{nm}) If $f: V \to W$ is a fiber space, dim $V = n$, dim $W = m$, and if F is the generic fiber, then

$$\kappa(V) \geqslant \kappa(W) + \kappa(F). \tag{13}$$

At present it has been proved for $m \leqslant n \leqslant 4$ and for $m = 1, n - 2$ or $n - 1$, and any n. It has various consequences for refinements of the classification, in particular for the case $\kappa(V) = -\infty$. The proofs of C_{nm} in the known cases use heavy machinery such as the variation of Hodge structures (IX, 83), Griffiths theory of period maps, and the theory of moduli of curves (IX, 12–14).

33. To progress in classification theory, the theory of moduli should be conveniently extended beyond curves and "polarized" abelian varieties (IX, 113), at least over the complex field. However, examples of families of surfaces are known for which "moduli spaces" cannot be defined, even if schemes are replaced by the more general "algebraic spaces" of M. Artin (VIII, 47). Such "algebraic spaces of moduli" have been defined only in special cases, for instance, surfaces of general type (IX, 25), K3 surfaces (IX, 23), and Enriques surfaces (IX, 23).

34. We finally should mention the extension of classification theory to *noncomplete* algebraic varieties, in particular the *quasi-projective* ones that are open subsets of projective varieties (for the Zariski topology). Such a variety can always be considered as a connected open set in a smooth compact algebraic variety $\overline{V}$, such that $\overline{V} - V$ is a divisor on $\overline{V}$ with normal crossings (IX, 46). Then properties of $\overline{V}$ are used to obtain a classification of V, independent of the compactification $\overline{V}$ chosen.

35. *B) Vector Bundles on Algebraic Varieties.* The study of an algebraic vector bundle E on an algebraic variety X is equivalent to the study of the corresponding sheaf of germs of sections $\mathcal{O}(E)$, defined by the condition that for any affine open set $U \subset X$, $\Gamma(U, \mathcal{O}(E))$ is isomorphic to the module $(\Gamma(U, \mathcal{O}_X))^r$ if r is the rank of E; such sheaves are called *locally free*. Most of the recent work on vector bundles on algebraic varieties is concentrated in three areas: vector bundles over $\mathbf{P}_n(\mathbf{C})$, stability of vector bundles, and vector bundles on complex noncomplete varieties.

36. I) *Vector Bundles over* $\mathbf{P}_n(\mathbf{C})$. For any quasi-projective (IX, 34) smooth algebraic variety X over $\mathbf{C}$, if $\text{Vect}^r_{\text{alg}}(X)$ and $\text{Vect}^r_{\text{an}}(X)$ are respectively the set of isomorphisms of algebraic and holomorphic vector bundles of rank r, there is a natural map

$$\text{Vect}^r_{\text{alg}}(X) \to \text{Vect}^r_{\text{an}}(X). \tag{14}$$

It has been known since the work of Serre (VIII, 24) that when X is *projective* the map (14) is bijective, and one may then apply to the study of algebraic vector bundles on X all the techniques of analytic geometry. This has been done for

$X = \mathbf{P}_n(\mathbf{C})$. A possible strategy consists of three steps: 1) classify the *topological* vector bundles over $\mathbf{P}_n$; 2) determine which topological vector bundles admit a holomorphic structure; and 3) for a given topological vector bundle, classify all possible holomorphic structures on it. Nothing more than partial results are known for this program, which is thus full of open problems.

37. (1) Concerning the first step in the preceding strategy, it is enough to consider the bundles of rank $r \leqslant n$. As $H^{2i}(\mathbf{P}_n; \mathbf{Z}) \simeq \mathbf{Z}$ for all $i \leqslant n$, one can identify the Chern classes $c_i(E)$ of a vector bundle to an integer. In 1966, Schwarzenberger noticed that the Chern classes of a topological vector bundle E over $\mathbf{P}_n$ satisfied the relations

(15) $$\sum_{i=1}^{r} \binom{\delta_i}{k} \in \mathbf{Z} \qquad \text{for} \qquad 2 \leqslant k < n$$

where the δ_i are related as usual to the total Chern class by

(16) $$c(E) = \prod_{i=1}^{n} (1 + \delta_i).$$

These conditions classify the bundles of rank $r = n$, but not those of rank $r < n$. Necessary and sufficient conditions for a pair (c_1, c_2) of integers to be the Chern classes of a complex vector bundle of *rank 2* over $\mathbf{P}_n$ have only been found for $n \leqslant 6$; in particular, it is known that conditions (15) are not sufficient for $n = 5$, $r = 2$. Furthermore, there may exist nonisomorphic topological vector bundles of rank 2 on $\mathbf{P}_n$ with the same pair (c_1, c_2) of Chern classes.

38. 2) With regard to the second step, there is no general theory for $n \geqslant 2$, but merely various methods of construction of algebraic vector bundles. For $n = 1$, Grothendieck proved in 1956 that all algebraic vector bundles on $\mathbf{P}_1$ have the form (VIII, 21)

(17) $$E = \mathcal{O}_{\mathbf{P}_1}(a_1) \oplus \mathcal{O}_{\mathbf{P}_1}(a_2) \oplus \cdots \oplus \mathcal{O}_{\mathbf{P}_1}(a_r)$$

for integers a_j. This leads to the idea of restricting a vector bundle E on $\mathbf{P}_n$ to a straight line L in $\mathbf{P}_n$, so that formula (17) applies to that restricted bundle $E|L$ (for integers a_j dependent on L). The lines L for which $E|L$ is not isomorphic to $E|L_0$, where L_0 is a general line, are called *jumping lines* for E; they have been studied in several special cases. In particular, results have been obtained for *uniform* bundles, that is, those for which there are no jumping lines.

In contrast with the case $n = 1$, for $n \geqslant 3$ there are *indecomposable* algebraic vector bundles on $\mathbf{P}_n$ of rank $n - 1$, and an indecomposable vector bundle of rank 2 (resp. 3) has been constructed on $\mathbf{P}_4$ (resp. $\mathbf{P}_5$).

39. Vector bundles of rank 2 have been the subject of many investigations. To a closed smooth subvariety Y of $\mathbf{P}_n$ (for $n \geqslant 3$) of codimension 2, one can associate an algebraic vector bundle E of rank 2, such that Y is the scheme of zeros of a holomorphic section of E; $c_2(E)$ is then the degree of Y, and $c_1(E) = k$ is such that

the $\mathcal{O}_Y$-Module associated to the normal bundle of Y in $\mathbf{P}_n$ (IX, 65) is isomorphic to $\mathcal{O}_X(k)|Y$. This construction yields many examples of algebraic vector bundles of rank 2; in particular one can show that *any* topological vector bundle of rank 2 on $\mathbf{P}_3$ may be given an algebraic structure. However, examples are known of topological vector bundles with no algebraic structure.

40. The third step in the strategy above (IX, 36) merges with the second area of the most active present research, namely:

II) *Stability of Vector Bundles.* One says an algebraic vector bundle E of rank r on $\mathbf{P}_n$ is *stable* if, for all locally free subsheaves $\mathscr{F}$ of rank $s < r$ of the locally free sheaf $\mathscr{E}$ associated to E, one has between the Chern classes the relation

$$\frac{c_1(\mathscr{F})}{s} < \frac{c_1(\mathscr{E})}{r} \tag{18}$$

(if the sign $<$ is replaced by $\leqslant$, E is *semi-stable*). For such bundles having fixed Chern classes $c_1, c_2, \ldots, c_r$, it is possible to define a "*space of moduli*," which is an analytic space in general. This has been particularly studied for $r = 2$ over $\mathbf{P}_2$ and $\mathbf{P}_3$, and explicitly described for some pairs (c_1, c_2). A remarkable and surprising fact is that stable bundles of rank 2 on $\mathbf{P}_3$ are closely related to the solution of the Yang-Mills equations in the modern theory of elementary particles.

It is possible to extend the concept of stable vector bundles on any algebraic variety X. When X is a surface of general type (IX, 25), the study of stable vector bundles of rank 2 on X leads to interesting results on algebraic curves contained in X; if their genus is bounded, they form a "limited" family.

41. III) *Vector Bundles on Noncomplete Varieties.* When a quasi-projective smooth variety X over $\mathbf{C}$ (IX, 34) is not projective (for instance an affine variety (VIII, 20)), examples show that the natural map (14) may be neither injective nor surjective. The main problem then is to determine additional properties of a holomorphic bundle E on X that imply it is algebraic. One gets such a property by considering on E a hermitian metric and its curvature; if (in a well-determined sense) this curvature does not increase too fast in the neighborhood of $\overline{X} - X$ (where $\overline{X}$ is a smooth completion of X), then E is algebraic. The proof, due to Cornalba and Griffiths, is analytic, using $\bar{\partial}$-cohomology, elliptic operators, and a generalization of the classical Nevanlinna theory.

42. *C) Group Actions on Algebraic Varieties.* Let X be an irreducible algebraic variety over an algebraically closed field k, G an algebraic group over k, and $(g, x) \mapsto g \,.\, x$ a *group action* of G on X (this means that the maps $(g, g') \mapsto g^{-1}g'$ of $G \times G$ into G, and $(g, x) \mapsto g \,.\, x$ of $G \times X$ into X are *morphisms* with the usual properties $e \,.\, x = x$ and $(g_1 g_2) \,.\, x = g_1 \,.\, (g_2 \,.\, x)$). If one excepts the theory of linear groups and of their homogeneous spaces (VIII, 25), the work done recently on such actions principally concerns the cases in which G is *commutative*, and more precisely the cases in which G is either an *abelian variety* (VII, 60) or an *algebraic torus* $T = (\mathbf{G}_m)^d$, where $\mathbf{G}_m$ is the algebraic group over k, such that its points in

any algebra A over k (VIII, 42) form the multiplicative group A^* of invertible elements of A. We shall consider only the work done on a type of action that has many surprising relations with apparently unconnected theories, the *toroidal embeddings*.

43. For a torus T, an *equivariant embedding* $j: T \to X$ is an open immersion such that $j(T)$ is dense in X, and the translations $(t, s) \mapsto ts$ in T extend to an action of T on X, $(t, x) \mapsto j(t).x$ (that is, $j(t).j(s) = j(ts)$). One first studies the case in which $X = \operatorname{Spec}(A)$ (VIII, 28) for a k-algebra A (so-called *torus embeddings*). Then $T = \operatorname{Spec}(k[M])$, where $k[M]$ is the algebra over k of the commutative group M of characters of T (a group isomorphic to $\mathbf{Z}^d$). The previous conditions on T and X are shown to be equivalent to the fact that $A \simeq k[S]$, the algebra over k of a *semi-group* $S \subset M$, finitely generated (as semi-group) and generating the group M. One can, by normalization of X (VII, 44), limit the investigation to the case in which S is "saturated," that is, the relation $ns \in S$ for an integer $n \neq 0$ implies $s \in S$. These semi-groups can be described geometrically as $S = \sigma \cap M$, where σ is a *polyhedral convex cone*, the intersection of a finite number of half spaces defined by $l_i(x) \geqslant 0$, where the l_i are linear forms on $\mathbf{R}^d$ with *rational* coefficients. One has thus a bijection $\sigma \mapsto X_\sigma = \operatorname{Spec}(k[\sigma \cap M])$ of the set of these cones on normal torus embeddings.

44. The *general* toroidal embeddings $T \to X$ for normal varieties X can then be obtained by gluing together a finite family (X_{σ_α}), where the σ_α are such that their polar cones $\check{\sigma}_\alpha$ in $(\mathbf{R}^d)^*$ form a conical polyhedral complex, also called a *fan*, that is, are such that $\check{\sigma}_\alpha \cap \check{\sigma}_\beta$ is a face of both $\check{\sigma}_\alpha$ and $\check{\sigma}_\beta$ if it is not empty. This description allows one to get very precise information on the singular points of X, on its topology, and on the $\mathcal{O}_X$-Modules.

45. Among the relations of toroidal embeddings with various parts of mathematics, one should first note the study of algebraic groups consisting of Cremona transformations of some $\mathbf{P}_n(k)$ (VI, 55 and 58) made by Demazure in 1970. It was this study that led Demazure to introduce for the first time the notion of toroidal embeddings, and with their help he could classify the groups he was considering. But the association of algebraic varieties to "fans" can also work in the opposite direction. Thus, the Hodge-Lefschetz theory (VII, 4–11) translates into quite nontrivial combinatorial properties of convex polytopes.

4. Singularities

46. *A) Resolution of Singularities.* One of the most spectacular recent results in algebraic geometry has been the proof by Hironaka, in 1964, of the resolution of singularities (VI, 32 and VII, 44) for algebraic varieties of *arbitrary* dimension over an algebraically closed field of *characteristic zero*. For such a variety X, there is a proper (VIII, 40) surjective morphism $f: X' \to X$, where X' is a *smooth* algebraic

variety, and an open dense subset U of X such that the restriction of f to $f^{-1}(U)$ is an isomorphism onto U, and $X - U$ contains all the singular points of X. In addition $f^{-1}(X - U)$ is a divisor on X′ (VIII, 3) having only "normal crossings." This means that locally $f^{-1}(X - U)$ is defined by an equation

(19) $$\xi_1^{a_1}\xi_2^{a_2}\ldots\xi_d^{a_d} = 0$$

where the ξ_j are local coordinates, and the a_j are integers > 0.

The proof, which occupies over 200 pages, is a gigantic multiple induction. More than anywhere else, in the theory of singularities, it is difficult to separate algebraic geometry and analytic geometry, much of the work being of *local* character. Hironaka's proof also applies to real analytic spaces, and he later extended it to holomorphic spaces. It should also be stressed that, although dealing with varieties over a field, he had to work at times with S-schemes, where S is the spectrum of a local ring (VIII, 31).

Over algebraically closed fields of characteristic $p > 0$, Abhyankar has proved the resolution of singularities for varieties of *dimension 2*, and more recently for varieties of *dimension 3*, provided $p > 5$.

47. *B) Isolated Singularities.* An n-dimensional holomorphic complex variety has real dimension $2n$, and may be studied from the point of view of algebraic topology; in particular, its topological structure in the neighborhood of a singular point may be examined. The simplest case is that of a hypersurface V in $\mathbf{C}^{n+1}$, having a local equation $f(z) = 0$, where f is a holomorphic function in the neighborhood of $O \in \mathbf{C}^{n+1}$, with $f(O) = 0$, and V having no singular point $\neq 0$ in a small neighborhood of O. The problem was first considered by Brauner, Burau, and Zariski for $n = 1$, then by Mumford for $n = 2$, and in 1966 Milnor attacked the general case.

One takes in $\mathbf{C}^{n+1} = \mathbf{R}^{2n+2}$ the intersection K_ε of V and of a sphere $S_\varepsilon : |z| = \varepsilon$ of small enough radius, which is a smooth $(2n - 1)$-dimensional manifold. When O is not a singular point of V, K_ε is an unknotted sphere in the $(2n + 1)$-dimensional manifold S_ε; but for $n = 1$ and relatively prime integers $p, q > 2$, when one takes for V the curve $z_1^p + z_2^q = 0$, K_ε is a "torus knot" in $\mathbf{S}_3$. In general, the map $\varphi : z \mapsto f(z)/|f(z)|$ of $S_\varepsilon - K_\varepsilon$ into the unit circle $\mathbf{U}$ defines $S_\varepsilon - K_\varepsilon$ as a *fiber bundle* over $\mathbf{U}$. The fibers $F_\theta = \varphi^{-1}(e^{i\theta})$ are smooth parallelizable $2n$-dimensional manifolds, and have the homotopy type of a "wedge" of μ n-dimensional spheres (i.e. joined together by a single common point). The number μ is called the *Milnor number* of V at the point O. There are several other interpretations of μ, as for instance the multiplicity of the solution $z = 0$ of the equation $\operatorname{grad} f(z) = 0$, and it can also be brought in relation with the "Newton polyhedron," a generalization of the Newton polygon (III, 5) obtained by considering in $\mathbf{R}^{n+1}$ the points having as coordinates the exponents of the z_j in the monomials with nonzero coefficient in f.

48. Most papers on isolated singularities concern the case of *surfaces*; what really intervenes is only the *local ring* R at the singular point P, so that in fact the surface

may be replaced by the scheme $X = \mathrm{Spec}(R)$ for *any* 2-dimensional local ring R (which includes in particular the local rings at the isolated singular points of complex *analytic* surfaces). Most of the results are relative to the case in which R is an algebra over a field K, and the closed point P of X is *normal* (VII, 43). A morphism $\pi : X' \to X$ is a *resolution* of the scheme X if it is proper (VIII, 40) and birational, its restriction to $\pi^{-1}(X - P)$ is an isomorphism onto $X - P$, and finally X' has regular local rings (VII, 41) along the exceptional divisor $E = \pi^{-1}(P)$. Let E_j $(1 \leqslant j \leqslant r)$ be the irreducible components of E; E is connected, and when R is a local algebra over **C**, it was proved by Du Val in 1944 and Mumford in 1961 that the matrix $((E_i \, . \, E_j))$ formed by the intersection numbers (VI, 37) is *negative nondegenerate*.

49. The *genus* of X is by definition $\dim_K H^1(X', \mathcal{O}_{X'})$ (VIII, 10); it is independent of the choice of the resolution $X' \to X$. The singularities of genus 0 have been called *rational*, and have been much studied since 1960; but for $K = \mathbf{C}$, they had already been considered by Du Val in 1934 as "those which do not affect the conditions of adjunction." For a minimal resolution, the components E_j of E then exhibit the following properties:

1) the E_j are smooth rational curves;
2) for $i \neq j$, E_i and E_j are in "general position" (VI, 36) and $E_i \cap E_j \cap E_k = \varnothing$ for any 3 distinct indices;
3) there is no cycle $\{E_{j_1}, E_{j_2}, \dots, E_{j_s}, E_{j_1}\}$ such that any two consecutive E_{j_h} intersect.

50. One associates to the E_j a *weighted graph* having the E_j as vertices with weight $(E_j \, . \, E_j)$, and two vertices E_i, E_j being linked by an edge if and only if $(E_i \, . \, E_j) \neq 0$; the three preceding properties show that the graph associated to a rational singularity is always a *tree*. The most remarkable result is that for a rational singular point of *multiplicity 2*, all the possible weighted graphs are precisely the *Dynkin diagrams* of the theory of complex simple Lie algebras. When $K = \mathbf{C}$, only the graphs of types A_n, D_n, E_6, E_7, E_8 are possible; this had already been noticed by Du Val.

51. Rational isolated singularities occur in various contexts. For instance, they are the only possible singular points of the Del Pezzo surfaces (VI, 53), as shown by Du Val. Another example is given by the *quotient singularities*, that is, the singularities of the varieties of orbits X/G, where X is a smooth variety and G a group acting on X (IX, 42) and subject to various additional conditions.

52. *(C) Equisingularity.* Next to isolated singularities, most studies of singular points are relative to the notion of *equisingularity*, introduced by Zariski in 1965, and to which he has devoted a long series of papers. In its simplest form, there is an affine hypersurface V in $\mathbf{C}^n$, O is a singular point of V and W is a subvariety of V, of codimension 1, whose points are singular for V; but O is a simple point of W. A plane through a point P of W in a neighborhood of O, transversal to W, cuts V

along a curve Γ_P; *equisingularity at O along* W means that Γ_O and Γ_P, for $P \in W$ sufficiently close to O, are curves with "equivalent" singularities at O and P, respectively. One first therefore has to define "equivalent" singular points on algebraic plane curves; there are several ways of doing this, which fortunately give the same notion.

The study of equisingularity when codim $W > 1$ appears to be a thorny one. Zariski has proposed a definition by induction on codim W; the question is linked to similar problems in which the algebraic structure is weakened to a differential structure ("Whitney stratification"). Very active work is now proceeding on these problems.

5. Divisors and cycles

53. *A) Algebraic Cycles.* We have seen (VII, 28) how one can define on a *smooth* projective irreducible variety X of dimension n over an arbitrary algebraically closed field k, the additive group $C^r(X)$ of *algebraic cycles of codimension* r (i.e. of dimension $n-r$) for $1 \leqslant r \leqslant n$. We have also seen how the group of *divisors* $C^1(X)$ has been thoroughly studied (VII, 52–62). More recently, work has begun in the much more difficult theory of the $C^r(X)$ for $r \geqslant 2$. The subgroup $C^r_{\text{alg}}(X)$ of cycles *algebraically equivalent to 0* has been defined (VII, 32). The notion of linear equivalence of divisors (VI, 43) is generalized for $r \geqslant 2$ to *rational equivalence* of cycles: two cycles Z_1, Z_2 of codimension r are rationally equivalent if there is on $X \times k$ a cycle Z, intersecting each fiber $X \times \{t\}$ in a subvariety of codimension r, and such that $Z_1 = Z \cap (X \times \{0\})$ and $Z_2 = Z \cap (X \times \{1\})$. The cycles that are *rationally equivalent to* 0 constitute a subgroup $C^r_{\text{rat}}(X)$ of $C^r_{\text{alg}}(X)$. It can again be proved that, for any two cycles Z, Z' that intersect properly (VII, 29), the class of the cycle $Z \,.\, Z'$ in $A_{\bullet}(X) = \bigoplus_{r \geqslant 0} C^r(X)/C^r_{\text{rat}}(X)$ only depends on the classes of Z and Z′; and for arbitrary cycles Z, Z', there always are cycles Z_1, Z'_1 respectively rationally equivalent to Z and Z′ and intersecting properly (Chow's "moving lemma"). This gives on $A_{\bullet}(X)$ the structure of a graded associative and commutative ring, called the *Chow ring* of X.

54. The existence of Chow coordinates (VII, 31) shows that $C^r(X)/C^r_{\text{alg}}(X)$ is always an at most *countable* group. Its study is helped by the introduction of some other equivalence relations (cf. VII, 52). *Numerical equivalence* is defined as in (VII, 21), and $C^r_{\text{num}}(X)$ is the subgroup of $C^r(X)$ consisting of cycles numerically equivalent to 0. If $C^r_{\tau}(X)$ is the subgroup of $C^r(X)$ consisting of cycles that are *torsion equivalent to 0* (i.e. some integral multiple of which is algebraically equivalent to 0), one has $C^r_{\text{num}} \supset C^r_{\tau} \supset C^r_{\text{alg}}$.

55. Finally, if $k = \mathbf{C}$, to each algebraic cycle Z of codimension r is associated its homology class in $H_{2n-2r}(X; \mathbf{C})$, hence by Poincaré duality, a cohomology class $\gamma_r(Z)$ in $H^{2r}(X; \mathbf{C})$, which gives a homomorphism $\gamma_r : C^r(X) \to H^{2r}(X; \mathbf{C})$. The inverse image $\gamma_r^{-1}(0)$ is the subgroup of cycles *homologically equivalent to 0*, written $C^r_{\text{hom}}(X)$.

When k is an arbitrary algebraically closed field, there are still canonical homomorphisms $\gamma_r : C^r(X) \to H_l^{2r}(X)$ for l-adic cohomology (IX, 85) (resp. $C^r(X) \to H^{2r}(X/W)$ for crystalline cohomology (IX, 89)); in every case, $\gamma^\bullet = (\gamma_r)_{r \geqslant 0}$ transforms intersections of cycles into cup products.

Summing up, one has a sequence of subgroups

$$C^r \supset C^r_{\text{num}} \supset C^r_{\text{hom}} \supset C^r_\tau \supset C^r_{\text{alg}} \supset C^r_{\text{rat}}. \tag{20}$$

56. For $r = 1$ (divisors) we have seen that $C^1_{\text{num}} = C^1_{\text{hom}} = C^1_\tau$, and C^1/C^1_{num} is a finitely generated free commutative group (VII, 53). In general, it is still true that C^r/C^r_{num} is free and finitely generated. For $k = \mathbf{C}$, one has $C^r_{\text{num}} = C^r_{\text{hom}}$ for $r = 1, 2, n-1$, and n; but Griffiths has given examples in which $C^2_{\text{hom}} \neq C^2_\tau$. Examples are also known for which $C^r_{\text{hom}} \neq C^r_\tau$ for arbitrary fields k. It is not known (for $r > 1$) if $C^r_{\text{hom}}/C^r_\tau$ is always finitely generated, nor if $C^r_\tau/C^r_{\text{alg}}$ is finite.

57. For $k = \mathbf{C}$ and $r > 1$, one would hope that $C^r_{\text{alg}}/C^r_{\text{rat}}$ could be identified with an abelian variety, for instance the $J_{n-r}(X)$ defined by Weil (VII, 58). Lieberman has shown that the natural image of $C^r_{\text{alg}}(X)$ in $J_{n-r}(X)$ (VII, 58) is indeed an abelian variety, but not necessarily isomorphic to $C^r_{\text{alg}}(X)/C^r_{\text{rat}}(X)$. For the special case in which X is a three-dimensional projective smooth variety with $p_g = q = 0$, it has been shown that $C^2_{\text{alg}}(X)/C^2_{\text{rat}}(X)$ is isomorphic to $J_1(X)$, and this has been generalized to any field k.

58. The study of C^n_{alg} goes back to Severi for $k = \mathbf{C}$. Since cycles of dimension 0 are combinations $Z = \sum_i n_i P_i$ for $n_i \in \mathbf{Z} - \{0\}$, and P_i are distinct points of X, there is a natural homomorphism $C^n_{\text{alg}}(X) \to A(X)$ into the Albanese variety (VII, 59), which is surjective and has a kernel containing $C^n_{\text{rat}}(X)$; but a deeper study has shown that $C^n_{\text{alg}}(X)/C^n_{\text{rat}}(X)$ is much too big. The *degree* $\sum_i n_i$ defines a homomorphism $C^n(X) \to \mathbf{Z}$, whose kernel contains $C^n_{\text{rat}}(X)$, hence there is a homomorphism $B(X) = C^n(X)/C^n_{\text{rat}}(X) \to \mathbf{Z}$, whose kernel is $B_0(X) = C^n_{\text{alg}}(X)/C^n_{\text{rat}}(X)$. *Effective* cycles in $C^n(X)$ are those for which all $n_i > 0$. The effective cycles of dimension 0 whose degree m is > 0 may be considered as points in the mth symmetric product $S^m(X)$ (variety of orbits of the symmetric group $\mathfrak{S}_m$ acting on X^m). We thus get a map $\gamma: S^m(X) \to B(X)$. It can be proved that the fibers of that map in $S^m(X)$ are *countable* unions of closed irreducible subsets of $S^m(X)$. The *dimension* of such a union is by definition the *maximum* of the dimensions of its irreducible parts. This enables one to attach to γ a number

$$d_m = \dim S^m(X) - \mu_m \tag{21}$$

where μ_m is the *minimum* of the dimensions of the fibers of γ. Mumford was the first to show that for a surface X with geometric genus $p_g > 0$, the sequence (d_m) is *unbounded.* His result was greatly generalized by Roitman, who obtained, for all smooth projective irreducible varieties over $\mathbf{C}$, the expression

$$d_m = m \,.\, d(X) + j(X) \qquad \text{for } m \geqslant m_0 \text{ large enough,} \tag{22}$$

where $d(X)$ and $j(X)$ are integers $\geqslant 0$ such that $d(X) \leqslant \dim X$. Furthermore, he showed that if, for some integer $q \geqslant 2$, $H^0(X, \Omega_X^q) \neq 0$, then $d(X) \geqslant q$. Finally examples show that $d(X)$ can be *any* integer d such that $0 \leqslant d \leqslant \dim X$ and $d \neq 1$.

These results prove that $B_0(X)$ *cannot* be the set of rational points on an abelian variety for $d(X) > 0$.

59. Another direction of research concerns the *smoothing problem* for cycles. In general, an r-dimensional cycle Y will be the formal sum of subvarieties of X that will have *singular points*. Is it possible, for one of the equivalence relations described above, to find an r-dimensional cycle Y_1 equivalent to Y and that is the formal sum of *smooth* subvarieties? For rational equivalence, positive results have been obtained by Hironaka and Kleiman by different methods: if $r \leqslant \min(3, (n-1)/2)$, Y can be smoothed, and if $r \leqslant (n-2)/2$, then $(n-r-1)!Y$ can be smoothed. But there is an example of a cycle of dimension 7 in a projective variety of dimension 9, which is not smoothable.

60. *B) Ample Sheaves.* For a complete (VII, 38) Serre variety X (which may have singularities) over an algebraically closed field k, one may define, as in (VIII, 2), the notion of *Cartier divisor* (or simply *divisor*): one has simply to replace "meromorphic function" by "rational function" and "holomorphic function" by "regular function" (i.e. functions belonging to the ring A(V) (VIII, 27) for some affine open set V). The topology is of course the Zariski topology. This notion is different from the notion (called *Weil divisor*) extending to arbitrary fields k the definition of (VI, 43); however, both notions coincide when X is smooth. As in (VIII, 7), to each Cartier divisor D is associated a locally free $\mathcal{O}_X$-Module $\mathcal{O}_X(D)$ of rank 1 (IX, 35); conversely, such a sheaf (also called an *invertible* $\mathcal{O}_X$-Module) has the form $\mathcal{O}_X(D)$, where D is only determined up to linear equivalence (VIII, 4). For any closed subvariety Y of X, the restriction to Y of an invertible $\mathcal{O}_X$-Module $\mathcal{O}_X(D)$ is an invertible $\mathcal{O}_Y$-Module written $\mathcal{O}_Y(D)$.

61. Even when X is not smooth, one can define intersection numbers for a system $(D_j)_{1 \leqslant j \leqslant s}$ of divisors on X and a closed subvariety Y of dimension s, by the formula

$$(23)\quad ((D_1 D_2 \cdots D_s) \,.\, Y) = \sum_{t=0}^{s} (-1)^{s-t} \sum_{j_1 < j_2 < \cdots < j_t} \chi(\mathcal{O}_Y(D_{j_1} + D_{j_2} + \cdots + D_{j_t}))$$

where χ is the Euler-Poincaré characteristic of a coherent $\mathcal{O}_Y$-Module defined in (VIII, 23, formula (36)).

62. One says a divisor D (or the invertible $\mathcal{O}_X$-Module $\mathcal{O}_X(D)$) is *ample* if there is an embedding of X in a projective space P and an integer $n > 0$ such that nD is linearly equivalent to a section of X by a hyperplane of P (generalization of (VI, 25)); one says then that the divisor nD is *very ample.* There is a remarkable "numerical" criterion for ample divisors proved by Nakai and Moishezon: D is ample if, for every $s \geqslant 1$ and every closed subvariety Y of dimension s, one has

$(D^s . Y) > 0$. Examples show that it is not enough to assume that $(D . C) > 0$ for every closed curve C (i.e. the criterion restricted to $s = 1$).

63. This criterion has enabled Kleiman to prove a conjecture of Chevalley: if X is complete and smooth, there exists an embedding of X into a projective space if and only if every finite set of points of X is contained in some *affine* open subset of X. There are examples, due to Nagata and Hironaka, of smooth complete algebraic varieties that do not satisfy that criterion, hence are not projective.

64. For any divisor D on a projective variety X, the Euler-Poincaré characteristic $\chi(mD)$ (VIII, 12) is equal to a polynomial in m for large values of m. Matsusaka has recently proved the following result: for any integer-valued polynomial $P(m)$, there is an integer m_0 such that, for *every* smooth projective complex variety X and *every* ample divisor D on X, the relation $\chi(mD) = P(m)$ for all large enough m implies that mD is *very ample* for $m \geqslant m_0$. This generalizes the tricanonical projective embedding of curves (VI, 26) and the fact that 5Δ is very ample for some surfaces of general type (IX, 25).

65. The notion of ample divisor (or ample line bundle, or ample invertible $\mathcal{O}_X$-Module) has been generalized by Hartshorne to vector bundles (or locally free $\mathcal{O}_X$-Modules) of finite rank > 1 (IX, 35). Among the several equivalent definitions, the simplest is the following one: $\mathscr{E}$ is ample if on the projective scheme $\mathbf{P}(\mathscr{E})$ (VIII, 43), the invertible Module $\mathcal{O}_{\mathbf{P}(\mathscr{E})}(1)$ is ample. The most remarkable example of ample vector bundle is the tangent bundle of the projective space $\mathbf{P}_n(k)$; this property implies that for any smooth subvariety Y of $\mathbf{P}_n(k)$, the normal bundle of Y (quotient of the tangent bundle of $\mathbf{P}_n(k)$ restricted to Y by the tangent bundle of Y) is also ample. On the other hand, the tangent bundle of a hypersurface Y in $\mathbf{P}_n(k)$, of degree $d > 1$, is not ample.

66. It was conjectured for some time that, conversely, every smooth complete irreducible n-dimensional variety having an ample tangent bundle was isomorphic to $\mathbf{P}_n(k)$. This was first proved for small values of n only, and finally, in 1978, S. Mori succeeded in proving the conjecture for all n. His proof is very interesting for two reasons: first, it uses the existence (proved by Grothendieck) of a scheme Mor(C, X) "parametrizing" the morphisms of a projective smooth curve C into a projective smooth variety X. Second, the proof is done at first for an algebraically closed field k *of characteristic* $p > 0$, and then the case of characteristic 0 is *deduced* from the case of characteristic $p > 0$, using "reduction modulo p" (VIII, 37). We shall again encounter this remarkable method of proof (IX, 145).

67. *C) The Riemann-Roch Theorem on Singular Varieties.* As divisors and their associated invertible Modules may be defined on an arbitrary projective (smooth or not) variety X over an arbitrary algebraically closed field k (IX, 60), one may ask if there is again a "Riemann-Roch-Hirzebruch" formula generalizing the expression of $\chi(D)$ (VIII, 13, formula (24)) when X has arbitrary singularities.

68. For curves, this problem was solved by Rosenlicht in 1953. Only divisors D on the curve X, whose support does not contain any singular point of X, are considered; for these divisors, one has

$$\chi(\mathrm{D}) = \deg \mathrm{D} + 1 - \pi \tag{24}$$

where π is the "arithmetic genus" already considered by M. Noether for "ordinary" singularities (VI, 18). To define it in general, one considers for each singular point Q, the local ring $\mathcal{O}_{\mathrm{Q}}$ of that point and its integral closure $\mathcal{O}'_{\mathrm{Q}}$; the quotient $\mathcal{O}'_{\mathrm{Q}}/\mathcal{O}_{\mathrm{Q}}$ is a finite dimensional vector space over k, and one takes

$$\pi = g + \sum_{\mathrm{Q}} \dim_k(\mathcal{O}'_{\mathrm{Q}}/\mathcal{O}_{\mathrm{Q}}) \tag{25}$$

where g is the genus of X (VII, 44). The proof of (24) is done by considering the (smooth) normalization X′ of X (VII, 44) and applying the Riemann-Roch theorem to X′.

69. The generalization of the Riemann-Roch-Hirzebruch formula for any dimension was more difficult, and was only accomplished in 1973 by Baum, Fulton, and McPherson. For simplicity we first assume that $k = \mathbf{C}$, so that for any connected projective (smooth or not) variety X, the singular homology $\mathrm{H}_{\bullet}(\mathrm{X}; \mathbf{Q})$ and singular cohomology $\mathrm{H}^{\bullet}(\mathrm{X}; \mathbf{Q})$ over the rational field $\mathbf{Q}$ are defined. There is then a "cap product," defined in algebraic topology, as a bilinear map

$$\frown : \mathrm{H}^p(\mathrm{X}; \mathbf{Q}) \times \mathrm{H}_q(\mathrm{X}; \mathbf{Q}) \to \mathrm{H}_{q-p}(\mathrm{X}; \mathbf{Q}). \tag{26}$$

70. To understand the generalization, it is first necessary to go into some detail about Grothendieck's formulation of the theorem for a *smooth* variety M over **C** (although, of course, his goal was the generalization to an arbitrary algebraically closed field k (VIII, 23)). Let $\mathrm{Td}(\mathrm{M}) \in \mathrm{H}^{\bullet}(\mathrm{M}; \mathbf{Q})$ be the *Todd class* in cohomology, defined by the Todd polynomial (VIII, 13). On the other hand, if $\mathscr{E}$ is a locally free $\mathcal{O}_{\mathrm{M}}$-Module and $c(\mathscr{E})$ its total Chern class, the *Chern character* $\mathrm{ch}\,(\mathscr{E}) \in \mathrm{H}^{\bullet}(\mathrm{M}; \mathbf{Q})$ is defined by

$$\mathrm{ch}(\mathscr{E}) = \mathrm{rk}(\mathscr{E}) + \sum_i (e^{\delta_i} - 1) \tag{27}$$

where the δ_i are defined as usual by (16). Then the Riemann-Roch-Hirzebruch formula may be written

$$\chi(\mathscr{E}) = \langle \mathrm{ch}(\mathscr{E}) \smile \mathrm{Td}(\mathrm{M}), \mathrm{M} \rangle \tag{28}$$

or equivalently

$$\chi(\mathscr{E}) = (\mathrm{ch}(\mathscr{E}) \frown \tau(\mathrm{M}))_0 \tag{29}$$

where the right side is the part of degree 0 of the *homology class* $\mathrm{ch}(\mathscr{E}) \frown \tau(\mathrm{M})$, and

$$\tau(\mathrm{M}) = \mathrm{Td}(\mathrm{M}) \frown [\mathrm{M}] \tag{30}$$

where [M] is the fundamental homology class of M.

71. The generalization of (29) to *any* connected projective variety X is based on a convenient definition of the homology class $\tau(X)$, which reduces to (30) when X is smooth. In fact one defines $\tau(\mathscr{F})$ for *all coherent* $\mathscr{O}_X$-*Modules* $\mathscr{F}$, and then one takes $\tau(X) = \tau(\mathscr{O}_X)$. This is obtained by the following steps:

1) Embed X as a closed subvariety in a smooth variety M.
2) Form a resolution

(31) $$\mathscr{E}_{\bullet} : 0 \to \mathscr{E}_r \to \mathscr{E}_{r-1} \to \cdots \to \mathscr{E}_1 \to \mathscr{F} \to 0$$

where $\mathscr{F}$ is extended by 0 in M − X, the $\mathscr{E}_j$ are locally free $\mathscr{O}_M$-*Modules*, and the sequence (31) is *exact on* M − X; then the *homology class*

(32) $$\mathrm{ch}_X^M \mathscr{E}_{\bullet} = \left(\sum_i (-1)^i \mathrm{ch}(\mathscr{E}_i)\right) \frown [M]$$

in $H_{\bullet}(M; \mathbf{Q})$, belongs to the image of $H_{\bullet}(X; \mathbf{Q})$.

3) This implies that

(33) $$\mathrm{Td}(M) \frown \mathrm{ch}_X^M(\mathscr{E}_{\bullet})$$

may be considered as a *homology class in* $H_{\bullet}(X; \mathbf{Q})$.

4) Prove that this class is *independent* of the choice of the resolution $\mathscr{E}_{\bullet}$ and of the choice of the embedding $X \to M$.

It is then this class that is taken as $\tau(\mathscr{F})$ by definition, and one proves it has the expected properties, and thus

(34) $$\chi(\mathscr{E}) = (\mathrm{ch}(\mathscr{E}) \frown \tau(X))_0$$

for all locally free $\mathscr{O}_X$-Modules $\mathscr{E}$. Examples show that the previous definition (30) cannot be generalized directly, for it may happen that $\tau(X)$ is not equal to any homology class $\xi \frown [X]$ for any cohomology class ξ.

72. In this generalization, it is possible to replace the homology group $H_{\bullet}(X; \mathbf{Q})$ by the *Chow ring* $A_{\bullet}(X) \otimes_{\mathbf{Z}} \mathbf{Q}$ (IX, 53); of course, we have to define an auxiliary "cohomology ring" $A^{\bullet}_{\mathbf{Q}}$ and a "cap product"

(35) $$A^p(X) \times A_q(X) \to A_{q-p}(X)$$

having the usual properties. First, when Y is a *smooth* n-dimensional variety over *any* algebraically closed field k, take $A^q Y = A_{n-q} Y$. Then, if X is any quasi-projective variety, consider all morphisms $f: X \to Y$ into a smooth variety Y, and order these morphisms into a *direct system* by composition of morphisms. Then define

(36) $$A^{\bullet}X = \varinjlim A^{\bullet}Y$$

for this system. This means that $A^{\bullet}X$ is the quotient of the disjoint union of the $A^{\bullet}Y$, modulo the equivalence relation generated by the relations $y = g^* y'$, where $y \in A^{\bullet}Y, y' \in A^{\bullet}Y'$, and $g \circ f = f'$, where $f: X \to Y$ and $f': X \to Y'$ are morphisms into smooth varieties Y, Y', One makes $A^{\bullet}X$ into a ring by using the cartesian product $Y_1 \times Y_2$ to add or multiply classes in $A^{\bullet}\mathbf{Y}_1$ and $A^{\bullet}Y_2$. The cap product

(35) is obtained by considering the intersection $x._f y$, as defined by Serre (VII, 30), which belongs to $A_{q-p}X$ when $x \in A_p X$ and $y \in A^q Y$, f being a morphism $X \to Y$ into a smooth variety.

Finally, "Chern classes" belonging to $A^{\bullet}X$ are defined similarly. If $\mathscr{E}$ is a locally free $\mathcal{O}_X$-Module, there is a smooth variety Y, a morphism $f: X \to Y$, and a locally free $\mathcal{O}_Y$-Module $\mathscr{F}$ such that $f^*\mathscr{F} = \mathscr{E}$. One proves that the image of $c(\mathscr{F}) \in A^{\bullet}Y$ in $A^{\bullet}X$ does not depend on the choices of f and $\mathscr{F}$, and it is the total Chern class $c(\mathscr{E}) \in A^{\bullet}X$ by definition.

The argument sketched above to prove (34) can then be done without change; it gives back in particular the Grothendieck-Washnitzer previous generalizations for smooth varieties over arbitrary algebraically closed fields.

73. *D) Duality.* The duality theorem of Serre, proved for vector bundles (or, equivalently for locally free sheaves) on a projective smooth n-dimensional variety X (VIII, 22) was the subject of far-reaching extensions by Grothendieck and his school. Serre's theorem can be put in the following form: for any locally free $\mathcal{O}_X$-Module $\mathscr{E}$, there exists a canonical bilinear form

$$H^p(\mathscr{E}) \times H^{n-p}(\mathscr{E}^{\vee} \otimes \Omega_X^n) \to k \tag{37}$$

which is *nondegenerate*, hence identifies each factor in the product to the dual of the other one.

74. Grothendieck's first generalization concerns an arbitrary *closed irreducible reduced n-dimensional subscheme* of a projective space $P = \mathbf{P}_N(k)$, and an arbitrary *coherent* $\mathcal{O}_X$*-Module* $\mathscr{F}$, which replaces $\mathscr{E}$ in (37).

The bilinear map that replaces (37) is a particular case of the general "Cartier-Yoneda pairing" in homological algebra:

$$R^p T(F) \times \operatorname{Ext}^q(F, G) \to R^{p+q}T(G) \tag{38}$$

where F and G belong to an abelian category $\boldsymbol{C}$ with enough injectives, $T: \boldsymbol{C} \to \boldsymbol{C}'$ is a left exact additive functor into another abelian category $\boldsymbol{C}'$, and R^pT is the pth derived functor of T.

In the application to schemes, $\boldsymbol{C}$ is the category of coherent $\mathcal{O}_X$-Modules, $\boldsymbol{C}'$ the category of k-vector spaces, and T the functor $\mathscr{F} \mapsto H^0(\mathscr{F})$ (sections of $\mathscr{F}$ over X (VIII, 8)). Finally G is the *dualizing sheaf* ω_X, which is Ω_X^n when X is smooth, and in general is given by

$$\omega_X = \operatorname{Ext}^{N-n}_{\mathcal{O}_P}(\mathcal{O}_P, \Omega_P^N) \qquad \text{restricted to X;} \tag{39}$$

the right side of (39) is the sheaf with fibers $\operatorname{Ext}^{N-n}_{\mathcal{O}_s}(\mathcal{O}_s, \Omega_s^N)$ at every point $s \in P$, which already appears in Serre's paper (VIII, 21). The pairing thus takes the form

$$H^p(\mathscr{F}) \times \operatorname{Ext}^{n-p}(\mathscr{F}, \omega_X) \to H^n(\omega_X) \xrightarrow{\sim} k. \tag{40}$$

But this bilinear form is not always nondegenerate: additional conditions are required for that on the local rings $\mathcal{O}_x$. These are always satisfied when X is smooth, and then (40) reduces to Serre's mapping (37).

75. This result was later generalized further by eliminating the projective space P. The natural injection $X \to P$ is replaced by an arbitrary morphism of finite type $f: X \to Y$, in the spirit of Grothendieck's formulation of the Riemann-Roch theorem (VIII, 23). One then uses the formalism of *derived categories* of J. L. Verdier; its principal virtue is an improvement of the machinery of spectral sequences, but it is too technical to be described here.

A generalization in a different direction, for $k = \mathbf{C}$, has been given by Mebkhout, for sheaves of differential operators.

76. *E) Schubert Calculus.* One of the famous Hilbert problems (number 15) was "to establish rigorously the results published by Schubert on enumerative geometry." We have given in (VII, 23) an example of a particular case of the general problem consisting in finding the number of linear projective varieties of dimension d (called *d-planes*, for short) in $\mathbf{P}_n(\mathbf{C})$, satisfying given algebraic conditions. A large part of Schubert's papers was devoted to this problem, and it has attracted new investigations since 1930.

Since the d-planes of $\mathbf{P}_n$ may be considered as points of the *grassmannian* $\mathbf{G}_{d,n}(\mathbf{C})$, the problem can be interpreted as finding the *number of intersections* of algebraic subvarieties of $\mathbf{G}_{d,n}$. With the homological interpretation of intersections (VI, 37), the problem splits into: 1) the determination of an explicit basis for the *homology* **Z**-*module* $H_\bullet(\mathbf{G}_{d,n}; \mathbf{Z})$ and of its multiplication table (VI, 36); and 2) for each specific algebraic subvariety of $\mathbf{G}_{d,n}$ under consideration, find the explicit expression of its homology class in the basis of the homology **Z**-module $H_\bullet(\mathbf{G}_{d,n}; \mathbf{Z})$.

77. The first problem was substantially solved by Schubert's introduction of special subvarieties of $\mathbf{G}_{d,n}$: let $A_0 \subset A_1 \subset \cdots \subset A_d$ be a strictly increasing sequence in which A_j is an a_j-plane, and consider the set of d-planes L satisfying the "Schubert condition":

$$\dim(A_j \cap L) \geqslant j \quad \text{for } 0 \leqslant j \leqslant d. \tag{41}$$

This set, written $[A_0, A_1, \ldots, A_d]$ is called a *Schubert variety*; it is a closed algebraic variety of dimension $\sum_j (a_j - j)$.

It was first proved by Ehresmann in 1934 that the homology classes $(a_0, a_1, \ldots, a_d)$ of the Schubert varieties formed a basis for the **Z**-module $H_\bullet(\mathbf{G}_{d,n}; \mathbf{Z})$, which was self-dual for the pairing

$$(x, y) \mapsto (x \,.\, y) \tag{42}$$

(intersection number, taken equal to 0 when x and y do not have complementary dimensions).

The multiplication table of that basis had in fact been previously determined by Pieri and Giambelli around 1900. Giambelli had described a system of "special" Schubert varieties, which are generators of the *homology ring* $H_\bullet(\mathbf{G}_{d,n}; \mathbf{Z})$. The methods of Pieri and Giambelli are linked to the theory of symmetric functions.

78. To express the homology class of an algebraic subvariety W of $\mathbf{G}_{d,n}$, as linear combination of the Schubert classes $(a_0, a_1, \ldots, a_d)$, we only need to find the intersection number of W with a Schubert variety $[A_0, A_1, \ldots, A_d]$ in general position; this means finding the number of d-planes L satisfying at the same time conditions (41) and the conditions that define W. Innumerable computations of such numbers were done by Schubert and his contemporaries such as Salmon, Cayley, M. Noether, Halphen, C. Segre, and Zeuthen. As examples, one may give the number of common straight lines contained in two "congruences" of straight lines in $\mathbf{P}_3$ (IV, 9), or the number of straight lines contained in the intersection of two quadrics in $\mathbf{P}_4$, or the number of multisecants to a curve in some $\mathbf{P}_n$, and so on.

79. Schubert classes are closely linked to Chern classes, since the latter can be defined as pullbacks of cohomology classes of grassmannians. In fact, it was by using the Schubert calculus to compute the number of d-planes tangent to a d-dimensional variety and satisfying certain conditions, that Severi (and later Todd and Eger) was led to introduce the "canonical classes" later found to be the duals of Chern classes of the tangent bundle (VIII, 13). Chern classes also play an important part in modern research on the enumerative theory of singularities of morphisms, justifying and extending the rich store of formulas bequeathed by classical geometers for curves (VI, 24) and surfaces.

80. But Schubert did not limit himself to problems involving families of linear varieties. He also incorporated in his work the problems on families of plane conics treated by Chasles and De Jonquières, and in particular the determination of the number of conics tangent to 5 conics in general position (IV, 13). To proceed as above, it seems natural to replace the grassmannians by the variety V of *all* conics (including those that "degenerate" into 2 straight lines or 1 double line) identified with $\mathbf{P}_5(\mathbf{C})$ (IV, 9). If W_C is the subvariety of V consisting of conics tangent to a given one C, the problem amounts to computing (with respect to a suitable basis) the homology class w_C of W_C in $H_\bullet(V; \mathbf{Z})$, and then computing w_C^5 in that ring. It is fairly easy to see that W_C is a hypersurface of degree 6 in $\mathbf{P}_5$, hence a straight application of Bezout's theorem would give the number $6^5 = 7776$, which is too large. The explanation lies in the fact that when C varies, *all* hypersurfaces W_C contain the Veronese surface (VI, 54), hence Bezout's theorem is not applicable. By considering the case in which C degenerates in 2 distinct lines, and invoking the "principle of conservation of number," Schubert obtained the formula $w_C = 2(w_P + w_L)$, where w_P (resp. w_L) is the homology class of the variety W_P (resp. W_L) of conics containing a given point P (resp. tangent to a given line L), a result equivalent to Chasles's expression $2(\mu + \nu)$ in his theory of "characteristics" (IV, 13). The computation of w_C^5 is then reduced to those of $w_P^\alpha w_L^\beta$ for $\alpha + \beta = 5$, which are easy.

81. The justification of Chasles's and Schubert's procedure was taken up by several mathematicians. The key idea was provided by Study in 1886: a conic,

considered as the "envelope" of its tangents, may also "degenerate" into 2 points. Therefore, Study proposed to assign to a nondegenerate conic, not the point u in $\mathbf{P}_5$ having as homogeneous coordinates the coefficients of its equation, but the pair (u, v) in $\mathbf{P}_5 \times \mathbf{P}_5$, where the homogeneous coordinates of v are the coefficients of the *dual* conic. The variety V should then be defined as the *closure* of the set of those points (u, v). Using homology theory, it is then not hard to determine the homology group $H_2(V; \mathbf{Z})$ and to compute the class of W_C in that group, justifying the Schubert formula.

82. Going beyond Chasles's results, Schubert and several of his contemporaries tackled similar more complicated problems for other curves and surfaces, in particular quadrics and twisted cubic curves in $\mathbf{P}_3$. Schubert thus published two spectacular results: the number of quadrics tangent to 9 given ones in general position is 666,841,088, and the number of twisted cubics tangent to 12 quadrics in general position is 5,819,539,783,680. Up to now, no satisfactory justification of these theorems has been proposed using ideas similar to Study's, namely taking into consideration all possible "degeneracies." It cannot therefore be said that Hilbert's problem is completely solved.

6. Topology of algebraic varieties

83. *A) The Hodge Theory.* We have seen (VII, 4–10) that on a projective smooth irreducible algebraic complex variety, Hodge's theory gives profound information on its topology. But it does not apply to an algebraic variety over **C** that is not projective or singular. However, Deligne has found that in general the cohomology algebra $H^{\bullet}(X; \mathbf{C})$ of such a variety X still carries canonical filtrations, defining what he calls a *mixed Hodge structure*, that we cannot describe here. This theory partakes as much of analytic geometry as of algebraic geometry, and is closely related to the classical theory of singular points of linear differential equations in the complex domain. Much geometric information can be deduced from a general study of the "variation" of mixed Hodge structures depending on a parameter.

84. It follows from Dolbeault's formula (VIII, 11, formula (12)) that, for a complex projective smooth variety X, the groups $H^j(\Omega^r_X)$ have a purely algebraic definition, since the sheaves Ω^r_X are defined algebraically (VIII, 22). It is therefore also possible to define them for a projective smooth variety over an arbitrary algebraically closed field k. However, if k has characteristic $p > 0$, most results concerning the Hodge numbers $h^{r,s} = \dim_k H^s(\Omega^r_X)$ and the Albanese variety (VII, 10 and VIII, 11) lose their validity. There are examples for which $h^{0,1} \neq h^{1,0}$, or $h^{0,1} > \dim_k A$ or $h^{1,0} > \dim_k A$, or $h^{0,1} + h^{1,0} > 2 \dim_k A$. Even the relation $\dim_k A = \dim_k H^1(\mathcal{O}_X)$ is not always true. Here, a deeper analysis, due to Grothendieck and Mumford, has clarified the situation, and incidentally gives

a striking example of the usefulness of considering nilpotent elements in the local rings of a variety (VIII, 38). The problem starts from unsuccessful attempts of the Italian geometers to prove the relation $q = h^{1,0}$ for a surface (VI, 48). They had used an argument resting on the fact that, for some "continuous (nonlinear) systems" of curves on a surface (VII, 52), the "characteristic series" on a generic curve of the system (defined as in (VIII, 17), but where $|C|$ must be replaced by curves of the system "infinitely close" to C) is *complete* (VI, 22). In the language of schemes, it turns out that this condition is equivalent to the fact that the Picard scheme $\mathbf{Pic}_{X/k}$ (VIII, 43) of the surface X is *reduced* (i.e. its local rings have no nilpotent elements $\neq 0$). This is true for all projective surfaces when k has characteristic 0, but there are counterexamples for k of characteristic $p > 0$, and Mumford has obtained necessary and sufficient condition on the cohomology of X to ensure that $\mathbf{Pic}_{X/k}$ is reduced.

The other "pathologies" of the Hodge numbers in characteristic $p > 0$ have not yet found a satisfactory explanation.

85. *B) The Various Cohomologies.* Around 1963, Grothendieck used his path-breaking ideas on "Grothendieck topologies" and sheaves for that "topology" (VIII, 46) to introduce, for a Serre variety X over an *arbitrary* algebraically closed field k (of any characteristic), cohomology spaces over fields of *characteristic* 0 (even if k has characteristic $p > 0$). If l is a prime number *distinct* from the characteristic of k, then, for the *étale* "topology," the cohomology groups $H^j(X_{et}, \mathbf{Z}/l^\nu\mathbf{Z})$ are defined for any integer $\nu > 0$, and form an inverse system of groups; the groups

(43) $$H^i_l(X) = \varprojlim H^i(X_{et}, \mathbf{Z}/l^\nu\mathbf{Z})$$

are by definition the *l-adic cohomology groups* of X. They are modules over the ring $\mathbf{Z}_l$ of *l-adic integers*; hence the

(44) $$H^i_l(X; \mathbf{Q}_l) = H^i_l(X) \otimes_{\mathbf{Z}_l} \mathbf{Q}_l$$

are vector spaces over the field $\mathbf{Q}_l$ of *l-adic numbers* (of characteristic zero). It is also possible to define groups of l-adic cohomology and of l-adic "cohomology with proper supports" for sheaves $\mathscr{F}$ on a scheme X called "$\mathbf{Q}$-constructible sheaves," which it is not possible to define more precisely here. They are written, respectively, $H^i(X, \mathscr{F})$ and $H^i_c(X, \mathscr{F})$ if no confusion may arise (the prime number l being always distinct from the characteristics of the residual fields $\kappa(x)$ for $x \in X$). When $\mathscr{F}$ is the constant sheaf $\mathbf{Q}_l$ on X, $H^i(X, \mathbf{Q}_l)$ is the same as the group (44). When $f: X \to Y$ is a proper morphism (VIII, 40), one can also define "higher direct images" $R^i f_*(\mathscr{F})$ as in (VIII, 31).

86. For projective smooth connected varieties over k, the $\mathbf{Q}_l$-vector spaces $H^\bullet(X, \mathbf{Q}_l)$ have remarkable properties, quite similar to those of the classical cohomology $\mathbf{Q}$-vector spaces of smooth complex varieties. The $\mathbf{Z}_l$-modules $H^i_l(X)$ are finitely generated, and if $n = \dim X$, then $H^i_l(X) = 0$ for $i > 2n$; for the Albanese variety A of X, $2 \dim_k A = \dim_{\mathbf{Q}_l} H^1(X, \mathbf{Q}_l)$. When $k = \mathbf{C}$, there is a functorial isomorphism

(45) $$\mathrm{H}_l^i(\mathrm{X}) \xrightarrow{\sim} \mathrm{H}^i(\mathrm{X}^{\mathrm{an}};\mathbf{Z}) \otimes_{\mathbf{Z}} \mathbf{Z}_l$$

linking l-adic cohomology of X with the cohomology groups of the complex manifold X^{an} underlying X (VIII, 24); hence the knowledge of $\mathrm{H}^i(\mathrm{X}^{\mathrm{an}};\mathbf{Z})$ is equivalent to that of $\mathrm{H}_l^i(\mathrm{X})$ for *all* prime numbers l; (45) shows that the *rank* of the $\mathbf{Z}_l$-module $\mathrm{H}_l^i(\mathrm{X})$ is then independent of l.

87. Important properties of the $\mathrm{H}^i(\mathrm{X},\mathbf{Q}_l)$ for projective smooth varieties are first a "Künneth formula"

(46) $$\mathrm{H}^{\bullet}(\mathrm{X}_1,\mathbf{Q}_l) \otimes \mathrm{H}^{\bullet}(\mathrm{X}_2,\mathbf{Q}_l) \xrightarrow{\sim} \mathrm{H}^{\bullet}(\mathrm{X}_1 \times \mathrm{X}_2,\mathbf{Q}_l),$$

and second a "Poincaré duality" that has a formulation here slightly different from the classical case.

Let K be an algebraically closed field of characteristic 0; we denote by $\mathbf{Z}/l^n(1)$ the group of roots of unity in K of degree l^n. The map $x \mapsto x^{l^{m-n}}$ maps $\mathbf{Z}/l^m(1)$ onto $\mathbf{Z}/l^n(1)$ for $m > n$, and defines $(\mathbf{Z}/l^n(1))$ as an inverse system. One writes $\mathbf{Z}_l(1) = \varprojlim \mathbf{Z}/l^n(1)$; it is a $\mathbf{Z}_l$-module, isomorphic to $\mathbf{Z}_l$ (but there is no canonical isomorphism). Therefore, $\mathbf{Q}_l(1) = \mathbf{Z}_l(1) \otimes_{\mathbf{Z}_l} \mathbf{Q}_l$ is a $\mathbf{Q}_l$-vector space of dimension 1; $\mathbf{Q}_l(r)$ is defined as the tensor power $(\mathbf{Q}_l(1))^{\otimes r}$ and $\mathbf{Q}_l(-r) = (\mathbf{Q}_l(r))^{\vee}$ for $r > 0$. The "Poincaré duality" then results, first from the fact that there are "cup products" in l-adic cohomology, which are bilinear maps

(47) $$\smile : \mathrm{H}^p(\mathrm{X},\mathbf{Q}_l) \times \mathrm{H}^q(\mathrm{X},\mathbf{Q}_l) \to \mathrm{H}^{p+q}(\mathrm{X},\mathbf{Q}_l),$$

and a canonical isomorphism (for $n = \dim \mathrm{X}$)

(48) $$\mathrm{Tr} : \mathrm{H}^{2n}(\mathrm{X},\mathbf{Q}_l) \xrightarrow{\sim} \mathbf{Q}_l(-n),$$

so that one has a bilinear map $(x,y) \mapsto \mathrm{Tr}(x \smile y)$

(49) $$\mathrm{H}^j(\mathrm{X},\mathbf{Q}_l) \times \mathrm{H}^{2n-j}(\mathrm{X},\mathbf{Q}_l) \to \mathbf{Q}_l(-n),$$

which turns out to be a perfect pairing.

88. When k has characteristic $p > 0$, one may still take $l = p$ and define cohomology $\mathbf{Z}_p$-modules by relation (43), but they no longer have the preceding useful properties; in fact, Serre has shown by an example that it is not possible in this case to define a "reasonable" cohomology that would be a $\mathbf{Z}_p$-module or a $\mathbf{Q}_p$-vector space. For various reasons, it has transpired that a good ring of coefficients for a "good" p-adic cohomology would be the ring of *Witt vectors* $\mathrm{W}(k)$ over k, or its field of fractions K (which has characteristic 0). A first idea was to "lift" X to a proper and smooth scheme X′ over $\mathrm{W}(k)$ (i.e. X would be a "reduction" of X′ modulo the maximal ideal of $\mathrm{W}(k)$ (VIII, 37)), and to show that the De Rham cohomology over K of the "lifting" X′ is independent of the choice of that lifting. This was done by Monsky-Washnitzer and Lubkin. Unfortunately, such liftings do not always exist, and even if they do, the method can give no information on the p-torsion in cohomology.

89. Another method was proposed by Grothendieck under the name of *crystalline cohomology*, written $\mathrm{H}^{\bullet}(\mathrm{X}/\mathrm{W})$, and it was explored in detail by Berthelot. Its

definition uses, even more than the étale "topology," the full scope of the theory of *categories*, and the concepts of *site* and *topos*, which are the most general form of "Grothendieck topologies" and of the sheaves associated with them. For proper and smooth varieties X over k, the crystalline cohomology has all the good properties of the l-adic cohomology (for $l \neq p$) mentioned above.

90. Another method proposed by Grothendieck in 1966 is to use the concept of *hypercohomology* of a *complex* of modules, introduced by Cartan and Eilenberg in homological algebra. For instance, instead of defining the "Hodge cohomology" $\mathrm{H}^{\bullet}(\mathrm{X}) = (\mathrm{H}^{j}(\mathrm{X}))$ by $\mathrm{H}^{j}(\mathrm{X}) = \bigoplus_{r+s=j}\mathrm{H}^{s}(\Omega^{r}_{\mathrm{X}})$, one can define the "De Rham cohomology" as the hypercohomology $\mathbf{H}^{\bullet}(\Omega^{\bullet}_{\mathrm{X}})$. When $k = \mathbf{C}$, it is canonically isomorphic to the "topological" cohomology $\mathrm{H}^{\bullet}(\mathrm{X}; \mathbf{C})$; but when k has characteristic $p > 0$, it again fails to have the "good" properties. Hypercohomology may also be used to obtain a different definition of crystalline cohomology by embedding X in a smooth scheme Y over the ring $\mathrm{W}/p^{m}\mathrm{W}$, taking the hypercohomology of a suitable complex of sheaves on Y with support concentrated on X, and then taking the projective limit when m tends to $+\infty$.

91. Even if the various definitions of cohomology are not all as useful as the "good" ones, the investigation of their relations with the latter yields interesting geometric information. Such investigations have recently been pursued by S. Bloch in relation with K-theory and the Cartier theory of formal groups, and by L. Illusie using the hypercohomology of a complex defined by Deligne.

92. *C) The Algebraic Fundamental Group.* Ever since the invention of Riemann surfaces, the concept of *ramified covering* of an algebraic variety had been familiar (in a more or less precise way) to algebraic geometers, and the Italian school had used it not only for curves in $\mathbf{P}_2(\mathbf{C})$, but also for surfaces in $\mathbf{P}_3(\mathbf{C})$. It had a predominantly topological flavor, and when one associated a group to a covering, it was usually by the standard composition of loops and the use of homotopy.

93. After 1950, and the introduction of the new concepts of "abstract" algebraic geometry over an arbitrary field, it was necessary to give a purely algebraic foundation to these notions. The first steps in this direction were taken by Lang, Serre, and most actively by Abhyankar, who published many papers on this subject. For simplicity, we shall only consider *normal* irreducible varieties over a field k (not necessarily algebraically closed) (VII, 43). One says a *finite surjective* morphism $f: \mathrm{V}' \to \mathrm{V}$ turns an irreducible variety V′ into a *covering* of V (or that f itself is a covering). Then the field R(V′) of rational functions on V′ (VII, 37) is a *finite* extension of R(V) (when R(V) is identified to a subfield of R(V′) by means of f); the degree $[\mathrm{R}(\mathrm{V}') : \mathrm{R}(\mathrm{V})]$ is also called the *degree* of the covering. For each $x \in \mathrm{V}$, $f^{-1}(x)$ is a finite set of *at most n* points. If it has exactly n points, one says V′ is *unramified* over x. Moreover, V′ is unramified over *every* point of V if and only if f is an *étale* morphism (and then V′ is an *étale covering*) (VIII, 34); then one also says R(V′) is *unramified* over R(V).

94. Any covering V′ of V is determined up to isomorphism by the field extension $R(V) \subset K = R(V')$, for V′ is isomorphic to the variety obtained by the *normalization of* V *in* K (a generalization of the normalization process defined in (VII, 44): if $x \in X$ is a point then $\bigcap_{x' \in f^{-1}(x)} \mathcal{O}_{x'}$ is the subring of K consisting of the elements *integral* over $\mathcal{O}_x$).

95. If k is algebraically closed and V is smooth, the subvariety Δ of V where the covering V′ is ramified is called the *branch locus* of the covering; its irreducible components have codimension 1 in V ("purity of the branch locus").

96. The most interesting coverings V′ of V are those for which R(V′) is a Galois extension of R(V); they are called *galois coverings*. Then the Galois group $G = \mathrm{Gal}(R(V')/R(V))$ operates on V′, and V can be identified with the set of *orbits* of this action. Let C be an irreducible subvariety of V of codimension 1, and C′ an irreducible component of $f^{-1}(C)$ in the galois covering V′; the set of elements $\sigma \in G$ such that $\sigma(x') = x'$ for all $x' \in C'$ is a subgroup $I_{C'}$ of G, called the *inertia* subgroup of C′. One says that V′ is *tamely ramified over* C if all the groups $I_{C'}$ have an order prime to the characteristic exponent of k; they are then *cyclic* groups.

97. For a given algebraically closed extension Ω of R(V) and a given subvariety X of V, let K_X be the union of the finite Galois extensions $R(V') \subset \Omega$ of R(V) such that V′ is *unramified over* V − X and *tamely ramified* over every subvariety of V of codimension 1. Then the group $\mathrm{Gal}(K_X/R(V))$ is in general an infinite compact totally disconnected group, called the *tame algebraic fundamental group of* V *with respect to* X and written $\pi_1^{\mathrm{alg}}(V, X)$. When $X = \varnothing$, one just writes $\pi_1^{\mathrm{alg}}(V)$ and says it is the *algebraic fundamental group* of V. When $k = \mathbf{C}$ and V is smooth, $\pi_1^{\mathrm{alg}}(V, X)$ is the group obtained by *completion* of the topological fundamental group $\pi_1(V - X)$ for the topology defined by taking as neighborhoods of the identity all subgroups of finite index. For instance, if V is a smooth curve over **C** of genus $g \geqslant 1$, $\pi_1^{\mathrm{alg}}(V)$ has $2g$ *topological generators* s_j, t_j linked by the only relation

$$(s_1 t_1 s_1^{-1} t_1^{-1})(s_2 t_2 s_2^{-1} t_2^{-1}) \cdots (s_g t_g s_g^{-1} t_g^{-1}) = 1. \qquad (50)$$

An important example is the group $\pi_1^{\mathrm{alg}}(\mathbf{P}_1(k), S)$, where S is a finite subset of closed points of $\mathbf{P}_1$; one can then define it as the quotient of $\mathrm{Gal}(\overline{k(T)}/k(T))$ (where $\overline{k(T)}$ is an algebraic closure of $k(T) = R(\mathbf{P}_1)$) by the closed subgroup generated by the union of the groups of inertia of all places of the field $\overline{k(T)}$ lying over closed points of $\mathbf{P}_1 - S$.

98. For *unirational* smooth varieties V over **C** (IX, 29), $\pi_1^{\mathrm{alg}}(V) = 0$, and in particular $\pi_1^{\mathrm{alg}}(\mathbf{P}_n) = 0$. If C is an irreducible curve in the plane $\mathbf{P}_2(\mathbf{C})$ having as only singularities double points with distinct tangents, a theorem stated by Zariski, but with an incomplete proof, is that the group $\pi_1(\mathbf{P}_2 - C)$ is *commutative*. Deligne has recently given a complete proof for this theorem. It is also true that for any algebraically closed field k, the same assumptions on C imply that the tame fundamental group $\pi_1^{\mathrm{alg}}(\mathbf{P}_2(k), C)$ is commutative.

99. *D) Monodromy.* For the l-adic cohomology, the Lefschetz "monodromy" operator τ_j on the homology group $H_1(C_a; \mathbf{Z})$, described in (VI, 38), has an analog for a proper morphism $f: X \to S$, where X is smooth, connected, and of dimension $n+1$, and $S = \operatorname{Spec}(A)$, where A is a henselian discrete valuation ring with algebraically closed field of residues k. We denote by s the closed point of S (so that $\kappa(s) = k$), by η its generic point ($\kappa(\eta)$ being the field of fractions of A), and by $\bar{\eta}$ a "geometric point" above η (i.e. the spectrum of an algebraic closure $\overline{\kappa(\eta)}$ of $\kappa(\eta)$). One must think of S as corresponding to a small disk in $\mathbf{C}$ of center y_j in (VI, 38), s corresponding to the point y_j, $\bar{\eta}$ to a "general" point $t \neq y_j$ in the disk, and f to the projection pr_2.

100. It is assumed that the fiber $X_{\bar{\eta}}$ is smooth and X_s has a unique ordinary quadratic singularity; then there is a natural "specialization" homomorphism

$$\mathrm{sp}: H^i(X_s, \mathbf{Q}_l) \simeq H^i(X, \mathbf{Q}_l) \to H^i(X_{\bar{\eta}}, \mathbf{Q}_l). \tag{51}$$

It is an isomorphism for $i \neq n, n+1$, and there is an exact sequence

$$0 \to H^n(X_s, \mathbf{Q}_l) \xrightarrow{\mathrm{sp}} H^n(X_{\bar{\eta}}, \mathbf{Q}_l) \xrightarrow{x \mapsto \mathrm{Tr}(x \cup \delta)} \mathbf{Q}_l(m-n) \tag{52}$$
$$\to H^{n+1}(X_s, \mathbf{Q}_l) \to H^{n+1}(X_{\bar{\eta}}, \mathbf{Q}_l) \to 0.$$

Here $m = [\frac{n}{2}]$, and $\delta \in H^n(X_{\bar{\eta}}, \mathbf{Q}_l)(m)$ is the "evanescent cocycle" corresponding to the "evanescent cycle" δ_j of (VI, 38), and well defined up to sign. The inertial group $I = \mathrm{Gal}(\overline{\kappa(\eta)}/\kappa(\eta))$ acts on $X_{\bar{\eta}}$, hence on the $H^i(X_{\bar{\eta}}, \mathbf{Q}_l)$; for $i \neq n$, the action is trivial. For $i = n$, one has to distinguish two cases, according to the parity of n:

(1) If $n = 2m+1$ is odd, there is a canonical homomorphism $t_l: I \to \mathbf{Z}_l(1)$, and for $\sigma \in I$ and $x \in H^n(X_{\bar{\eta}}, \mathbf{Q}_l)$, one has

$$\sigma . x = x \pm t_l(\sigma) \mathrm{Tr}(x \smile \delta)\delta \qquad \text{and} \qquad \mathrm{Tr}(\delta \smile \delta) = 0 \tag{53}$$

(this is proved by "lifting" X to a scheme over a field of characteristic 0 and using the Picard-Lefschetz method).

(2) If $n = 2m$ is even and the characteristic of $\kappa(\eta)$ is $\neq 2$, there is a homomorphism $\varepsilon: I \to \{-1, 1\}$ and one has

$$\sigma . x = x \qquad \text{if} \qquad \varepsilon(\sigma) = 1, \tag{54}$$
$$\sigma . x = x \pm \mathrm{Tr}(x \smile \delta)\delta \qquad \text{for} \qquad \varepsilon(\sigma) = -1.$$

101. The previous results constitute the "local" Picard-Lefschetz method in l-adic cohomology. The "global" method used by Lefschetz to obtain the results of (VI, 38) on the homology of subvarieties extends also to the l-adic cohomology and is described as follows. Let $X \subset \mathbf{P}_N(k)$ be a smooth connected projective variety of dimension $n+1$; let M be a linear variety of codimension 2, and let $(H_t)_{t \in D}$ be the family of hyperplanes containing M, where $D = \mathbf{P}_1(k)$ is the projective line, so that $t \mapsto H_t$ is an injective morphism of D in the dual projective space $\mathbf{P}_N^\vee$. Let $\tilde{X} \subset X \times D$ be the set of pairs (x, t) such that $x \in H_t$, which is closed in $\mathbf{P}_N \times D$. One says the fibers $X_t = \mathrm{pr}_2^{-1}(t) = X \cap H_t$ constitute a *Lefschetz pencil* of n-dimensional subvarieties of X if one has the following properties:

1) M intersects X transversely (IX, 103); then $\tilde{X}$ is isomorphic to the blowing up of X along the $(n-1)$-dimensional variety $Z = M \cap X$ (the "axis" of the pencil).

2) There is a finite subset S of D and, for each $s \in S$, a single point x_s in the fiber X_s, such that pr_2 is smooth at all points of $\tilde{X}$ outside of the finite set of the x_s.

3) For each $s \in S$, x_s is an ordinary quadratic singular point of X_s.

From (1), it follows that the vector space $H^i(X, \mathbf{Q}_l)$ is naturally identified to a subspace of $H^i(\tilde{X}, \mathbf{Q}_l)$.

102. For every $s \in S$, one can apply the "local" monodromy results of (IX, 100) to the spectrum D_s of the henselization of the local ring $\mathcal{O}_s$ and the extension $\tilde{X} \times_D D_s \to D_s$ of pr_2 (VIII, 32). For a point $u \in D$, one may consider that the tame fundamental group $\pi_1^{alg}(D, S)$ (IX, 97) acts on the fiber X_u, hence on the $\mathbf{Q}_l$-vector spaces $H^i(X_u, \mathbf{Q}_l)$, the action being transported from the action of I defined in (IX, 100). Excluding the case when $p = 2$ and n is even, this action is trivial if $i \neq n$. For $i = n$, one considers the subspace E of $H^n(X_u, \mathbf{Q}_l)$ generated by the evanescent cocycles δ_s for all $s \in S$, and shows that E is *stable* under $\pi_1^{alg}(D, S)$ and that its orthogonal $E^\perp$ for the bilinear form $\mathrm{Tr}(x \smile y)$ is the subspace of the elements *invariant* under $\pi_1^{alg}(D, S)$. Thus a linear representation is

(55) $$\rho : \pi_1^{alg}(D, S) \to GL(E/(E \cap E^\perp)),$$

which is absolutely irreducible. If $E \neq 0$, the form $\mathrm{Tr}(x \smile y)$ induces on the vector space $E/(E \cap E^\perp)$ a bilinear nondegenerate form

(56) $$\psi : E/(E \cap E^\perp) \times E/(E \cap E^\perp) \to \mathbf{Q}_l(-n),$$

which is alternating if n is odd, symmetric if n is even. Furthermore, when n is odd, a theorem of Kazhdan-Margulis shows that the image $\rho(\pi_1^{alg}(D, S))$ is an *open subgroup* in the locally compact symplectic group $Sp(\psi)$ (for the l-adic topology of $\mathbf{Q}_l$).

103. *E) Cohomology and Subvarieties.* The Lefschetz theorem (also called "weak" or "easy" Lefschetz theorem) on the relation between the homology of a smooth projective compex variety and the homology of its smooth hyperplane sections (VI, 38) has been the source of many recent investigations.

A smooth subvariety Y of $\mathbf{P}_n(\mathbf{C})$ of dimension r is called a (strict) *complete intersection* if $Y = H_1 \cap H_2 \cap \cdots \cap H_{n-r}$, where the H_j are hypersurfaces that are nonsingular at the points of Y and have a *transversal* intersection (i.e. at each point of Y the tangent hyperplanes to the H_j are in general position). For such a subvariety Y, the Lefschetz theorems generalize as follows: the restriction map $H^i(\mathbf{P}_n; \mathbf{Z}) \to H^i(Y; \mathbf{Z})$ is bijective for $i < r$ and injective for $i = r$; Pic(Y) is a group isomorphic to $\mathbf{Z}$ and generated by $\mathcal{O}_Y(1)$ for $r \geqslant 3$ (VIII, 21) and $\pi_1(Y) = 0$ if $r \geqslant 2$.

104. An old problem going back to Kronecker and Cayley is to know if an irreducible subvariety of $\mathbf{P}_n(\mathbf{C})$ of dimension r is the intersection of $n - r$ hyper-

surfaces (with *no* condition on their tangent hyperplanes). In $\mathbf{P}_3(\mathbf{C})$, M. Kneser has shown that every irreducible (smooth or not) curve is the intersection of 3 surfaces, and recently Ferrand and Szpiro have shown that in *affine* space $\mathbf{C}^3$, every smooth curve is the intersection of 2 surfaces. In $\mathbf{P}_4(\mathbf{C})$, R. Hartshorne has given an example of an irreducible surface with a single singular point, which is not a complete intersection.

105. Conditions on the *degree* of a projective variety Y, which imply that Y is a complete intersection, have also been investigated. For instance, if Y is a smooth irreducible subvariety of $\mathbf{P}_n(\mathbf{C})$ of codimension 2 and of degree $\geqslant n/2$, then Y is a complete intersection. For given $d > 0$, there exists an integer $n_0(d)$ such that if $n \geqslant n_0$ and Y is a smooth subvariety of $\mathbf{P}_n(\mathbf{C})$ of degree d, not contained in any hyperplane, then Y is a complete intersection.

106. Recently, W. Barth has found that for a *smooth* subvariety Y of $\mathbf{P}_n(\mathbf{C})$ of dimension r, it is still possible to obtain information on the topology of Y even if Y is not a complete intersection. The map $H^i(\mathbf{P}_n; \mathbf{Z}) \to H^i(Y; \mathbf{Z})$ is still an isomorphism for $i \leqslant 2r - n$, $\pi_1(Y) = 0$ if $2r - n \geqslant 1$, and $\mathrm{Pic}(Y) \simeq \mathbf{Z}$ if $2r - n \geqslant 2$; Larsen has shown that the relative homotopy group $\pi_i(\mathbf{P}_n, Y) = 0$ if $i \leqslant 2r - n + 1$.

107. The cohomology of sheaves on *open* subsets $\mathbf{P}_n(k) - Y$ (where k is an algebraically closed field of arbitrary characteristic) has also been the subject of several papers. If Y is a closed subscheme of $\mathbf{P}_n(k)$ and $\mathbf{P}_n - Y$ is *affine*, Serre's criterion (VIII, 29) already shows that $H^i(\mathscr{F}) = 0$ for all $i > 0$ and any quasi-coherent sheaf $\mathscr{F}$. But there are plenty of open sets $\mathbf{P}_n(k) - Y$ that are not affine (for instance, the complement of two points in $\mathbf{P}_2$). However, there are some results on the $H^i(\mathscr{F})$ for some values of i and some kinds of schemes Y; for instance, $H^n(\mathscr{F}) = 0$ for all quasi-coherent sheaves if $Y \neq \varnothing$, and $H^{n-1}(\mathscr{F}) = 0$ if Y is connected and of dimension $\geqslant 1$. There are more complicated criteria for the other cohomology groups. A remarkable fact is that the proofs of these results are quite different for $k = \mathbf{C}$ and for fields of characteristic $p > 0$.

7. Abelian varieties

108. The general theory of algebraic geometry over an algebraically closed field k, of arbitrary characteristic, finds an attractive domain of application in the properties of abelian varieties over k (VII, 60–62). For instance, if X is an abelian variety of dimension g over k, the Hodge vector spaces $H^s(\Omega^r_X)$ over k (IX, 84) have a particularly simple structure:

(57) $$\dim_k H^1(\mathscr{O}_X) = \dim_k H^0(\Omega^1_X) = g$$

and there are canonical isomorphisms

(58) $$H^s(\Omega^r_X) \simeq \wedge^r(H^0(\Omega^1_X)) \otimes \wedge^s(H^1(\mathscr{O}_X)).$$

Algebraic equivalence of divisors (or invertible $\mathscr{O}_X$-Modules) on X is the same as numerical equivalence (IX, 54). For a divisor D, one has, with the notation of (IX, 61),

(59) $$\chi(\mathscr{O}_X(D)) = (D^g)/g!$$

109. The group underlying the *dual* $\hat{X}$ of the abelian variety X (VII, 62) is the subgroup $\mathrm{Pic}^0(X)$ of classes of divisors algebraically (or numerically) equivalent to 0.

For any $x \in X$, let t_x be the morphism $z \mapsto x + z$ of X onto itself; for any invertible $\mathscr{O}_X$-Module $\mathscr{L}$, let $\Lambda(\mathscr{L})(x)$ be the isomorphism class in Pic(X) of the invertible $\mathscr{O}_X$-Module $t_x^*\mathscr{L} \otimes \mathscr{L}^{-1}$. Then $\mu = \Lambda(\mathscr{L})$ is a group homomorphism of X into Pic(X), and one has $\Lambda(\mathscr{L}) = 0$ if and only if the class $\mathrm{cl}(\mathscr{L})$ in Pic(X) belongs to $\mathrm{Pic}^0(X)$.

In order that $\mathscr{L}$ or $\mathscr{L}^{-1}$ be ample (IX, 62), it is necessary and sufficient that the kernel of $\Lambda(\mathscr{L})$ be a finite subgroup. Then $\mu = \Lambda(\mathscr{L})$ is an *isogeny* (VII, 56) $X \to \hat{X}$, one has $H^i(\mathscr{L}) = 0$, for $i > 0$, μ has a degree (IX, 93) r^2, where $r = \chi(\mathscr{L})$, and $\mathscr{L}^{\otimes 3}$ is very ample. Conversely, every isogeny $\mu : X \to \hat{X}$ can be written $\mu = \Lambda(\mathscr{L})$ for an ample $\mathscr{O}_X$-Module $\mathscr{L}$, which is only determined up to numerical equivalence.

110. On the product $X \times \hat{X}$ there is a canonical line bundle P, called the *Poincaré line bundle*, such that its restriction to $X \times \{\alpha\}$ for any $\alpha \in \hat{X}$ is a line bundle on X whose corresponding invertible $\mathscr{O}_X$-Module has a class equal to α in Pic (X). If $\mathscr{P}$ is the invertible sheaf corresponding to P, then for an isogeny $\mu : X \to \hat{X}$, one defines canonically on X an invertible $\mathscr{O}_X$-Module $\mathscr{L}_X(\mu) = q^*(\mathscr{P})$, where $q = 1_X \times \mu$. If $\mu = \Lambda(\mathscr{L})$, $\mathscr{L}_X(\mu)$ is numerically equivalent to $\mathscr{L}^{\otimes 2}$.

111. If X and Y are two abelian varieties over k, the group Hom (X, Y) is a free **Z**-module of rank ρ such that $\rho \leqslant 4 \dim X . \dim Y$.

In many respects, the Tate module $T_l(X)$ of an abelian variety X (for a prime number l distinct from the characteristic of k) (VII, 61) plays a part similar to that of the Lie algebra of a Lie group. We have seen that to any homomorphism $u : X \to Y$ between abelian varieties, there corresponds a $\mathbf{Z}_l$-homomorphism $T_l(u) : T_l(X) \to T_l(Y)$; the natural map

(60) $$\mathbf{Z}_l \otimes_{\mathbf{Z}} \mathrm{Hom}(X, Y) \to \mathrm{Hom}_{\mathbf{Z}_l}(T_l(X), T_l(Y))$$

is injective. We shall see (IX, 146 and 153) that in some important cases it is bijective.

If f is an étale morphism of a variety X′ onto an abelian variety X, there is on X′ (up to a finite choice of origin) a uniquely determined structure of abelian variety, for which f is a homomorphism. The l-primary part of the algebraic fundamental group $\pi_1^{\mathrm{alg}}(X)$ (IX, 97) is $T_l(X)$.

112. There always exist on an abelian variety X ample invertible $\mathscr{O}_X$-Modules. In 1955, A. Weil introduced the notion of *polarized abelian varieties* that are pairs

$(X, \mathrm{cl}_a(\mathscr{L}))$, where $\mathrm{cl}_a(\mathscr{L})$ is the class of an ample invertible $\mathcal{O}_X$-Module $\mathscr{L}$, for algebraic equivalence. Such a class (or an element $\mathscr{L}$ of that class) is called a *polarization* of X, and $\chi(\mathscr{L})$ is the *degree* of the polarization. A *principal polarization* is a polarization of degree 1. The existence of such a polarization implies that X and $\hat{X}$ are isomorphic; for instance, the jacobian of a curve (VII, 57) is principally polarized by a canonical polarization. An *isomorphism*

$$\varphi : (X_1, \mathrm{cl}_a(\mathscr{L}_1)) \xrightarrow{\sim} (X_2, \mathrm{cl}_a(\mathscr{L}_2))$$

of polarized abelian varieties is an isomorphism $\varphi : X_1 \xrightarrow{\sim} X_2$ of abelian varieties, such that $\varphi^*(\mathscr{L}_2)$ is algebraically equivalent to $\mathscr{L}_1$. The automorphism group of a polarized abelian variety is *finite*, whereas the automorphism group of an abelian variety is infinite.

It is equivalent to define a polarization of degree d of an abelian variety X and an isogeny $\mu : X \to \hat{X}$ of degree d^2. The sheaf $\mathscr{L}_X(\mu)^{\otimes 3}$ is then very ample and defines a closed immersion $j : X \to \mathbf{P}_N$, with $N = 6^g d - 1$. The corresponding Hilbert polynomial of $j(X)$ (VIII, 23) is

$$P(m) = \chi(\mathscr{L}_X(\mu)^{\otimes 3m}) = 6^g d m^g. \tag{61}$$

113. Let Q be the projective scheme over k, which is the subscheme of the Hilbert scheme of $\mathbf{P}_N(k)$ (VIII, 43) whose points are the isomorphism classes of closed subvarieties of $\mathbf{P}_N$ having as Hilbert polynomial the polynomial (61). There is then a locally closed subscheme R of $Q \times \mathbf{P}_N$, on which the automorphism group PGL(N) operates canonically, and such that the orbit space may be identified (as a set) to the set $A_{g,d}$ of the isomorphism classes of *polarized* abelian varieties $(X, \mathscr{L})$ of dimension g and degree of polarization d. It may be shown that there is a natural structure of *quasi-projective scheme* on $A_{g,d}$. This scheme is called the *coarse moduli scheme* of polarized abelian varieties of dimension g and polarization degree d.

114. The concept of abelian variety can be generalized to that of *abelian S-scheme* for a noetherian scheme S. It is an S-group scheme X (VIII, 44) such that the morphism $f : X \to S$ is proper and smooth, and has geometric fibers $f^{-1}(s) \otimes_{\kappa(s)} \overline{\kappa(s)}$ that are abelian varieties. One can extend to schemes the concept of ample divisor, hence also the notion of *polarized* abelian S-scheme. Because a polarized abelian variety may have automorphisms other than the identity, the scheme $A_{g,d}$ does *not* represent the functor associating to a noetherian scheme S over k the set of isomorphism classes of polarized abelian S-schemes of dimension g and degree of polarization d.

115. To get a representable functor, at least over $\mathbf{C}$, one introduces on an abelian variety X of dimension g additional structures, the *rigidifications of level m* (m integer > 0). These are isomorphisms $\varepsilon : (\mathbf{Z}/m\mathbf{Z})^{2g} \xrightarrow{\sim} X_m$, where X_m is the kernel of the endomorphism $x \mapsto m \,.\, x$ of X, which is a subgroup of order m^{2g} (VII, 60). With an evident definition of isomorphism, the only automorphism of a polarized and rigidified abelian variety is the identity if m is large enough. Then there is a quasi-projective scheme $A_{g,d,m}$, parametrizing the abelian varieties of

given triple (g, d, m) of dimension, polarization degree, and level of rigidification; $A_{g,d}$ is the set of orbits of the action of the group $GL(2g, \mathbf{Z}/m\mathbf{Z})$ on $A_{g,d,m}$.

116. The classical theorem of Torelli says that two smooth projective curves over k are isomorphic if and only if their jacobians, equipped with the canonical principal polarization, are isomorphic as polarized abelian varieties. There is a space of moduli $M_{g,m}$ for smooth irreducible curves of genus $g \geqslant 2$ over k, equipped with a rigidification of level $m \geqslant 3$ of their jacobians; it is a smooth algebraic variety of dimension $3g - 3$ over k. There is a natural mapping $M_{g,m} \to A_{g,1,m}$ of $M_{g,m}$ into the moduli space of principally polarized abelian varieties of dimension g over k, equipped with a rigidification of level m. It follows from the Torelli theorem that this mapping is an injective morphism.

117. Let A, B, C be three *algebraic commutative group varieties* over a field k (VII, 60); then C is *an extension of* A *by* B if there are two group homomorphisms f, g:

$$0 \to B \xrightarrow{f} C \xrightarrow{g} A \to 0 \tag{62}$$

such that the sequence is *exact*, and in addition f and g are *separable morphisms* of algebraic varieties. Two extensions C, C′ of A by B are *isomorphic* if there is a diagram

$$0 \to B \begin{array}{c} \nearrow \\ \searrow \end{array} \begin{array}{c} C \\ \varphi \downarrow \\ C' \end{array} \begin{array}{c} \searrow \\ \nearrow \end{array} A \to 0$$

that is commutative and such that φ is an isomorphism of group varieties. The process used by Baer to define on the isomorphism classes of extensions a structure of commutative group when A, B, C are merely commutative groups applies as well when in addition A, B, C are group varieties; one writes again Ext(A, B) that group of isomorphism classes if there is no possible confusion.

The main results concern the case in which A is an abelian variety, and B is one of the groups $\mathbf{G}_m$ (IX, 42) or $\mathbf{G}_a$ (the algebraic group whose points in any k-algebra R form the additive group of R). One proves then that

$$\operatorname{Ext}(A, \mathbf{G}_m) \simeq \operatorname{Pic}^0 A \quad \text{and} \quad \operatorname{Ext}(A, \mathbf{G}_a) \simeq H^1(\mathcal{O}_A). \tag{63}$$

118. For $k = \mathbf{C}$, extensions of abelian varieties by tori $(\mathbf{G}_m)^r$ or vector groups $(\mathbf{G}_a)^r$ play an important part in the generalizations of the classical notions of "integrals of the second (resp. third) kind" introduced by Legendre for elliptic curves, and by Riemann for smooth algebraic curves of arbitrary genus (V, 12–13). The problem is of course what is meant by saying that a *rational* differential m-form on a smooth complex irreducible algebraic variety X (VII, 10) is of the second or third kind. For surfaces, this was done by Picard for second and third kind when $m = 1$, and for second kind when $m = 2$, in his very original pioneering work on algebraic surfaces. His results were extended by Lefschetz for higher dimensions, and more recently the subject has been taken up by Rosenlicht, Atiyah-Hodge, and Serre.

119. For arbitrary m, the definition now adopted for a *closed* rational m-form of the second kind on a smooth projective variety X is the following one: there is a divisor on X such that, if U is the open set, complement of its support, an m-form ω is of the second kind if it is holomorphic on U and if the cohomology class in $H^m(U^{an}, \mathbf{C})$ of its restriction $\omega|U$ is in the image of $H^m(X^{an}, \mathbf{C})$. This implies that locally, for the Zariski topology, one may write $\omega = \eta + d\tau$, where η is holomorphic and τ a rational $(m-1)$-form. However, this condition is sufficient for ω to be of the second kind if $m = 1$ or $m = 2$, but *not* for $m \geqslant 3$; this follows from Griffiths' example showing that $C^2_{\text{hom}} \neq C^2_\tau$ (IX, 56).

120. For $m = 1$, *any* closed rational 1-form ω on X is holomorphic in a Zariski open set that is the complement of the support of a divisor D. If the Y_i are the irreducible components of the support of D, and V is an affine open set in X, such that for each i, $f_i = 0$ is the equation of $Y_i \cap V$, there exists a rational function $g \in R(X)$ holomorphic in V and a holomorphic 1-form η on V, such that on V one has

$$\omega = \sum_i c_i \frac{df_i}{f_i} + dg + \eta. \tag{64}$$

The coefficient c_i is called the *residue* of ω along Y_i, and ω is a form of the second kind if and only if all residues $c_i = 0$. One says ω is a differential form of third kind if in (64) one may take $g = 0$. Weil has shown that a divisor $D = \sum_i c_i D_i$ on X (D_i irreducible subvarieties of codimension 1) is such that the c_i are the residues of a differential 1-form of the third kind, if and only if the Chern class of D in $H^2(X^{an}, \mathbf{C})$ is 0.

121. In 1949, Weil observed that one could define differential forms of the second (resp. third) kind on X as being obtained as follows: consider an extension G of the Albanese variety A of X by a vector group (resp. a torus), and the pullbacks $f^*\eta$ of a differential form η on G, invariant by translation. Serre has shown, more precisely, that for any divisor D on X, if U is the complement of the support Y of D, there are a well-determined extension G of A by a torus, and a rational function f of X into G that is a morphism in U (G and f depending on D) such that all 1-forms of the third kind whose divisor of residues has support contained in Y, can be written $f^*\eta$ for an invariant form η on G.

8. Diophantine geometry

122. Since 1960, the study of algebraic varieties defined over a finite field (VII, 47–51) or over a number field (VII, 35) has made considerable progress, and has been one of the main sources in the development of new concepts and methods in the theory of algebraic varieties over an arbitrary field and the theory of schemes.

123. *A) The Weil Conjectures.* In his work on the "Riemann hypothesis" for curves defined over a finite field $\mathbf{F}_q$, A. Weil had noticed that the definition of

the zêta function and its interpretation in terms of the fixed points of iterates of the Frobenius morphism (VII, 51) could be extended to projective varieties of any dimension. Let $|X_0|$ be a projective smooth irreducible Serre variety (VIII, 20) over the algebraic closure $\bar{\mathbf{F}}_q$ of $\mathbf{F}_q$, consisting of the *closed points* of a scheme X, algebraic and projective over $\mathrm{Spec}(\bar{\mathbf{F}}_q)$. We suppose that $X = X_0 \otimes_{\mathbf{F}_q} \bar{\mathbf{F}}_q$ is obtained by extension of the scalars from $\mathbf{F}_q$ to $\bar{\mathbf{F}}_q$ for a scheme X_0 over $\mathrm{Spec}(\mathbf{F}_q)$. As the finite extensions of $\mathbf{F}_q$ are the finite fields $\mathbf{F}_{q^m}$ for all $m \geqslant 1$, $|X_0|$ is the union of the sets $X_0(\mathbf{F}_{q^m})$ of "points of X_0 with values in $\mathrm{Spec}(\mathbf{F}_{q^m})$" (VIII, 42), that is, the points $x \in |X_0|$ such that $\kappa(x) \subseteq \mathbf{F}_{q^m}$.

124. We may assume $|X_0|$ is a subvariety of some projective space $\mathbf{P}_N(\bar{\mathbf{F}}_q)$. The *Frobenius automorphism* F of $\mathbf{P}_N(\bar{\mathbf{F}}_q)$ changes the projective coordinates of a point into their qth powers. It leaves globally invariant $|X_0|$, hence it is an automorphism of that variety. The fixed points of F in $|X_0|$ are the points of $X_0(\mathbf{F}_q)$, and the fixed points of F^m are the points of $X_0(\mathbf{F}_{q^m})$. If $N_m = \mathrm{Card}(X_0(\mathbf{F}_{q^m}))$ is the number of these fixed points, the *zêta function* $Z(X_0; t)$ of X_0 is defined by the formula

$$Z(X_0; t) = \exp\left(\sum_{m=1}^{\infty} N_m t^m/m\right) = \prod_{x \in |X_0|} (1 - t^{\deg(x)})^{-1} \tag{65}$$

where for each closed point $x \in |X_0|$, $\deg(x)$ is the degree $[\kappa(x) : \mathbf{F}_q]$.

125. In algebraic topology, the "algebraic" number of fixed points of a continuous "general" map f of a complex $\sum$ into itself is given by the *Lefschetz formula* (of which formula (28) of (VI, 41) is a particular case):

$$\sum_i (-1)^i \mathrm{Tr}(f_i) \tag{66}$$

where f_i is the endomorphism of the ith cohomology space $H^i(\sum; \mathbf{Q})$ deduced from f. Weil observed that if, for every smooth variety Y over $\bar{\mathbf{F}}_q$, "cohomology vector spaces" $H^i(Y)$ could be defined over a field K of *characteristic zero*, for which the Lefschetz formula still gave the number of fixed points of an automorphism of Y, then it would be possible to express the zêta function (65) by a formula (where $n = \dim|X_0|$)

$$Z(X_0; t) = \frac{P_1(t)\, P_3(t) \cdots P_{2n-1}(t)}{P_0(t)\, P_2(t) \cdots P_{2n}(t)} \tag{67}$$

where $P_i(t) = \det(1 - F_i t)$, F_i being the endomorphism of the vector space $H^i(|X_0|)$ deduced from the Frobenius automorphism F. Weil verified on simple examples such as the grassmannians, that (67) was correct, the degree of the polynomial P_i being equal to the *Betti number* R_i of an algebraic *complex* variety obtained by "lifting" $|X_0|$ in characteristic 0 (VIII, 41).

126. Led by these considerations and by his proof of the "Riemann hypothesis" for curves, Weil formulated in 1949 four conjectures on the zêta function (65). The first one was that $Z(X_0; t)$ is a *rational* function of t with coefficients in $\mathbf{Q}$; the second,

that there is a decomposition of $Z(X_0; t)$ of the form (67), where each P_i is a polynomial with *coefficients in* $\mathbf{Z}$ and constant term 1, and could be written

$$(68) \qquad P_i(t) = \prod_{j=1}^{b_i} (1 - \alpha_{ij} t)$$

where the "reciprocal zeros" of P_i satisfy

$$(69) \qquad |\alpha_{ij}| = q^{i/2}.$$

The third conjecture was that $Z(X_0; t)$ should satisfy a functional equation similar to F. K. Schmidt's one for curves (VII, 48, formula (24)), the map $\alpha \mapsto q^n/\alpha$ carrying bijectively the α_{ij} to the $\alpha_{2n-i,j}$.

Finally, if $|X_0|$ is obtained by "reduction modulo p" (VIII, 37) from a smooth complex variety Y, then the degree b_i of P_i should be the ith Betti number of Y.

127. In 1960 Dwork was able to prove directly (without using any cohomology) the rationality of $Z(X_0; t)$, even when $|X_0|$ is not smooth. In 1964–66 Grothendieck, M. Artin, and Verdier showed, for any prime number l distinct from the characteristic p of $\mathbf{F}_q (q = p^e)$ that for *l-adic cohomology* (IX, 85), one had the Lefschetz formula

$$(70) \qquad \mathrm{Card}(X_0(\mathbf{F}_{q^m})) = \sum_{i=0}^{2n} (-1)^i \mathrm{Tr}(F^m; H^i(\mathbf{X}, Q_l)) = \sum_{i,j} (-1)^i \beta_{ij}$$

where the β_{ij} are the eigenvalues of F acting on the cohomology $\mathbf{Q}_l$-space $H^i(X, \mathbf{Q}_l)$, and are therefore in the algebraic closure $\bar{\mathbf{Q}}_l$. This implied a decomposition similar to (67), P_i being replaced by the polynomial

$$(71) \qquad P_{i,l,X_0}(t) = \prod_{j=1}^{b_i} (1 - \beta_{ij} t) = \det(1 - tF; H^i(X, \mathbf{Q}_l)).$$

Furthermore, Poincaré duality in l-adic cohomology (IX, 87) implies the functional equation

$$(72) \qquad Z(X_0; \frac{1}{q^n t}) = \pm\, q^{n\chi/2} t^{\chi} Z(X_0; t)$$

with $\chi = \sum_{i=0}^{2n} (-1)^i \dim_{\mathbf{Q}_l} H^i(X; \mathbf{Q}_l)$. Finally, the relation between l-adic and ordinary cohomology for smooth projective complex varieties (IX, 86, formula (45)) showed that the degree b_i of P_{i,l,X_0} was given by the fourth Weil conjecture.

128. These results did not prove that the coefficients of the P_{i,l,X_0} were integers independent of l, nor that when $\bar{\mathbf{Q}}_l$ and $\mathbf{C}$ were identified by an isomorphism, the complex numbers α_{ij} corresponding to the β_{ij} satisfied (69). To establish these remaining and hardest parts of the Weil conjectures, Grothendieck analyzed Weil's proof of the "Riemann hypothesis" for $n = 1$, and enumerated a list of properties of the Lefschetz theory (VI, 38), which became known as the "standard conjectures." Grothendieck showed that if these standard conjectures could be proved for l-adic cohomology, all Weil conjectures would follow. However, even for $k = \mathbf{C}$, some of these properties have not yet been established for classical

cohomology, and those that have been proved rely essentially on Hodge theory. It came therefore as a surprise that in 1973 P. Deligne succeeded in proving all Weil conjectures by an entirely new method, bypassing the "standard conjectures," with the exception of the "weak Lefschetz theorem" (IX, 103 and VI, 38) whose validity in l-adic cohomology had been established by Grothendieck and Artin. It seems worthwhile to give a sketch of the main ideas of this remarkable proof, which consists of a series of very skillful reductions, finally making the problem manageable. The proof can be divided into two main parts.

129. *Part A)* The first reduction consists in restricting the proof to the case in which the dimension n of X is *even*, and to show that it is enough to prove, instead of (69), the inequalities

(73) $$|\alpha_{nj}| \leqslant q^{(n+1)/2}$$

(i.e. one has only to consider P_{n,l,X_0} instead of all the P_{i,l,X_0}). Indeed if the dimension n of X is arbitrary, one applies (73) to the product X^m where m is an *even* integer >0. By the Künneth formula, which is valid for l-adic cohomology (IX, 87), the β_{nk}^m are among the reciprocal roots of the polynomial P_{nm,l,X_0^m}, so that

$$|\alpha_{nk}^m| \leqslant q^{(nm+1)/2}.$$

Hence, by letting the even number m tend to $+\infty$, $|\alpha_{nk}| \leqslant q^{n/2}$. By the functional equation (72), q^n/β_{nk} is also a root of P_{n,l,X_0}, hence $|\alpha_{nk}| = q^{n/2}$, and (69) is proved for $i = n$. For $i < n$, one uses induction on $n - i$. If $|Z_0|$ is a smooth hyperplane section of $|X_0|$, the weak Lefschetz theorem shows that $H^i(X, \mathbf{Q}_l) \to H^i(Z, \mathbf{Q}_l)$ is injective, so that β_{ik} is a reciprocal root of P_{i,l,Z_0}, and one applies the induction hypothesis to $|Z_0|$. Finally, for $i > n$, one uses the functional equation (72), which shows that q^n/β_{ik} is a root of P_{2n-i,l,X_0}.

130. In order to make inductive proofs, the second idea is to bring to bear on the problem the Picard-Lefschetz *monodromy*, as extended to l-adic cohomology (IX, 101–102). However, one must not lose sight of the fact that the scheme X_0 is over $\mathrm{Spec}(\mathbf{F}_q)$, and that in the definition (65), it is X_0 and not $X = X_0 \otimes_{\mathbf{F}_q} \overline{\mathbf{F}}_q$ that intervenes. Therefore, one has to make sure that all the ingredients S, D, M, x_s, u in (IX, 101) are obtained from similar objects S_0, D_0, M_0, x_{s_0}, u_0 relative to X_0, by extending the base field $\mathbf{F}_q$ to its algebraic closure $\overline{\mathbf{F}}_q$. This is not always possible; however, it becomes so if one first extends $\mathbf{F}_q$ to a *finite* extension $\mathbf{F}_{q^r}$, works on $X' = X_0 \otimes_{\mathbf{F}_q} \mathbf{F}_{q^r}$, and replaces the embedding $X_0 \to \mathbf{P}_N$ by another one $X'_0 \to \mathbf{P}_{N'}$ with $N' = Nq^r$. Fortunately, this does not change the inequality (73), for q is replaced by q^r and β_{nj} by β_{nj}^r. The only point that requires some care concerns the comparison between the fundamental groups $\pi_1^{\mathrm{alg}}(D, S)$ and $\pi_1^{\mathrm{alg}}(D_0, S_0)$. Happily, this is again a simple matter, for there is an exact sequence

(74) $$0 \to \pi_1^{\mathrm{alg}}(D, S) \to \pi_1^{\mathrm{alg}}(D_0, S_0) \to \mathrm{Gal}(\overline{\mathbf{F}}_q/\mathbf{F}_q) \to 0$$

and $\mathrm{Gal}(\overline{\mathbf{F}}_q/\mathbf{F}_q)$ is the compact group $\hat{\mathbf{Z}}$ generated (topologically) by the Frobenius bijection $\xi \mapsto \xi^q$.

131. One also needs the extension of the Lefschetz formula (70) to cohomology of "constructible" sheaves $\mathscr{V}_0$ on the scheme X_0. For its inverse image $\mathscr{V}$ on X and every closed point x of X, the Frobenius automorphism F of X defines a homomorphism $F_x:\mathscr{V}_{F(x)} \to \mathscr{V}_x$, which is an endomorphism for $F(x) = x$, that is, $x \in |X_0|$. On the other hand, F operates on the l-adic cohomology with proper supports (IX, 85) $H_c^{\bullet}(X,\mathscr{V})$, and one has the formula

$$\sum_{x\in|X_0|} \mathrm{Tr}(F_x;\mathscr{V}_x) = \sum_i (-1)^i \mathrm{Tr}(F; H_c^i(X,\mathscr{V})). \tag{75}$$

If one considers the function generalizing (65) (and the Artin L-functions (VII, 61))

$$Z(X_0,\mathscr{V}_0;t) = \prod_{x\in|X_0|} (\det(1 - t^{\deg(x)}F_x;\mathscr{V}_x))^{-1}, \tag{76}$$

the formula (75) applied to all powers F^m gives the relation generalizing (67):

$$Z(X_0,\mathscr{V}_0;t) = \prod_i \det(1 - tF; H_c^i(X,\mathscr{V}))^{(-1)^{i+1}}. \tag{77}$$

132. With these preliminaries out of the way, and supposing the dimension n of X_0 *even*, a Lefschetz pencil on X allows one to suppose that there exists a projective morphism $f_0: X_0 \to D_0$ extending to a projective morphism $f: X \to D$, such that the fibers $X_u = f^{-1}(u)$ of closed points u of D are smooth $(n-1)$-dimensional varieties, except for u belonging to a finite subset S, and then X_u has a unique quadratic singularity.

The next step is to use the Leray spectral sequence for f that reduces the study of the eigenvalues of F operating on $H^n(X, \mathbf{Q}_l)$ to the study of the eigenvalues of F operating on the three cohomology groups:

$$H^0(D,\mathscr{R}^n), \qquad H^1(D,\mathscr{R}^{n-1}), \qquad H^2(D,\mathscr{R}^{n-2}) \tag{78}$$

where

$$\mathscr{R}^s = R^s f_*(\mathbf{Q}_l). \tag{79}$$

This is because D has dimension 1, hence all other terms E_2^{pq} of the sequence, such that $p + q = n$, are 0.

133. Since the "axis" Z of the Lefschetz pencil has the form $Z_0 \otimes_{\mathbf{F}_q} \overline{\mathbf{F}}_q$, where $|Z_0|$ is a projective smooth $(n-2)$-dimensional variety, it is possible to use induction on the even integers to dispose of $H^2(D,\mathscr{R}^{n-2})$. It is isomorphic to $H^{n-2}(X_u, \mathbf{Q}_l)(-1)$, and the latter, by the weak Lefschetz theorem and Poincaré duality, is contained in $H^{n-2}(Z, \mathbf{Q}_l)(-1)$.

To dispose similarly of $H^0(D, \mathscr{R}^n)$, one has to introduce the subspace E of evanescent cocycles in $H^n(X_u, \mathbf{Q}_l)$ (IX, 102). When $E \neq 0$, $H^0(D, \mathscr{R}^n)$ is identified to $H^n(X_u,\mathbf{Q}_l)$, and one has a surjective "Gysin homomorphism" $H^{n-2}(Z, \mathbf{Q}_l)(-1) \to H^n(X_u, \mathbf{Q}_l)$, which again allows use of induction. When $E = 0$, the exact sequence (52) of (IX, 100) is used.

The proof concerning $H^1(D, \mathscr{R}^{n-1})$ is more difficult, and this is the goal of part B.

134. *Part B* This part is first devoted to the proof of a *fundamental lemma* that abstracts the preceding situation in the following way. One has an affine curve U_0 over $\mathbf{F}_q$, smooth and absolutely irreducible, and one writes $U = U_0 \otimes_{\mathbf{F}_q} \overline{\mathbf{F}}_q$. On U_0 is given a locally constant $\mathbf{Q}_l$-constructible sheaf $\mathscr{V}_0$ whose image $\mathscr{V}$ in U has fibers $\mathscr{V}_u$ on closed points u of U, which are $\mathbf{Q}_l$-vector spaces on which the fundamental group $\pi_1^{\text{alg}}(U_0)$ operates, so that one has a linear representation

(80) $$\rho_0 : \pi_1^{\text{alg}}(U_0) \to \text{GL}(\mathscr{V}_u).$$

There is an exact sequence similar to (74):

(81) $$0 \to \pi_1^{\text{alg}}(U) \to \pi_1^{\text{alg}}(U_0) \to \hat{\mathbf{Z}} \to 0.$$

The *assumptions* of the fundamental lemma are:

(i) There is on $\mathscr{V}_0$ an alternating nondegenerate bilinear form

(82) $$\psi : \mathscr{V}_0 \times \mathscr{V}_0 \to \mathbf{Q}_l(-a) \qquad \text{for some } a \in \mathbf{Z}.$$

(ii) The image by ρ_0 of the subgroup $\pi_1^{\text{alg}}(U)$ in $\text{GL}(\mathscr{V}_u)$ is an open subgroup of the symplectic group $\text{Sp}(\mathscr{V}_u, \varphi_u)$ for the $\mathbf{Q}_l$-vector space $\mathscr{V}_u$.

(iii) For every $x \in |U_0|$, the polynomial $\det(1 - tF_x; \mathscr{V}_x)$ has rational coefficients.

The *conclusion* is that, for every $x \in |U_0|$, the eigenvalues of F_x in $\mathscr{V}_x$ are algebraic numbers whose conjugates all have absolute values $q_x^{a/2}$, where $q_x = q^{\deg(x)}$.

135. The proof starts by applying formula (77) of (IX, 131) to the tensor product $\mathscr{V}_0^{\otimes k}$ for an arbitrary integer $k > 0$. Let $Z_k(t) = Z(U_0, \mathscr{V}_0^{\otimes k}; t)$. Then on one hand

(83) $$Z_k(t) = \prod_{x \in |U_0|} 1/\det(1 - t^{\deg(x)} F_x; \mathscr{V}_x^{\otimes k});$$

on the other hand, since U_0 is affine, $H_c^0(U, \mathscr{V}^{\otimes k}) = 0$ so that

(84) $$Z_k(t) = P_k^1(t)/P_k^2(t),$$

where $P_k^i(t) = \det(1 - tF; H_c^i(U, \mathscr{V}^{\otimes k}))$ for $i = 1, 2$.

The fact that U_0 is affine also implies that

(85) $$H_c^2(U, \mathscr{V}^{\otimes k}) \simeq (\mathscr{V}_u^{\otimes k})_\pi(-1),$$

where $(\mathscr{V}_u^{\otimes k})_\pi$ is the largest quotient of $\mathscr{V}_u^{\otimes k}$ on which $\pi_1^{\text{alg}}(U)$ operates trivially. Due to assumption (ii), this is also the largest quotient on which the whole symplectic group $\text{Sp}(\mathscr{V}_u, \psi_u)$ operates trivially.

136. Now suppose $k = 2h$ is *even*. Classical invariant theory shows that linear forms on $(\mathscr{V}_u^{\otimes 2h})_\pi$ are the same thing as $2h$-linear forms on $\mathscr{V}_u$ invariant by $\text{Sp}(\mathscr{V}_u, \Psi_u)$, and these are sums of $2h$-linear forms

$$(v_1, \ldots, v_{2h}) \mapsto \prod_\alpha \psi_u(v_{i_\alpha}, v_{j_\alpha}),$$

where the $\{i_\alpha, j_\alpha\}$ form a partition of the set $\{1, 2, \ldots, 2h\}$ in parts such that $i_\alpha < j_\alpha$. The values of these forms are in $\mathbf{Q}_l(-ha)$ by assumption (i). Taking a maximal linearly independent set of such forms, one gets an isomorphism

(86) $$H_c^2(U, \mathscr{V}^{\otimes 2h}) \simeq \mathbf{Q}_l(-ha-1)^N \qquad \text{for some } N > 0.$$

Hence, as F acts on $\mathbf{Q}_l(m)$ by multiplication by q^{-m},

(87) $$P_{2h}^2(t) = (1 - q^{ha+1}t)^N.$$

Thus $Z_{2h}(t)$ has only poles of absolute value $\leqslant 1/q^{ha+1}$. But

$$\mathrm{Tr}(F_x^m; \mathscr{V}_x^{\otimes 2h}) = (\mathrm{Tr}(F_x^m; \mathscr{V}_x))^{2h}$$

is *positive*, and *rational* by assumption (iii). This implies that each power series on the right side of (83) has *positive* coefficients, hence has a radius of convergence *at least* as great as that of $Z_{2h}(t)$. Now if α is an eigenvalue of F_x on $\mathscr{V}_x$, α^{2h} is an eigenvalue of F_x on $\mathscr{V}_x^{\otimes 2h}$. Hence, for any α' conjugate to α, $1/\alpha'^{2h/\deg(x)}$ is a pole of

$$1/\det(1 - t^{\deg(x)} F_x; \mathscr{V}_x^{\otimes 2h}),$$

which shows that

$$|\alpha'| \leqslant q_x^{a/2 + 1/2h}$$

and, since h is arbitrarily great, $|\alpha'| \leqslant q_x^{a/2}$. But, due to the existence of ψ, $q_x^a \alpha^{-1}$ is also an eigenvalue, so one has $|q_x^a \alpha'^{-1}| \leqslant q_x^{a/2}$, and thus $|\alpha'| \geqslant q_x^{a/2}$.

137. This lemma implies as a corollary that the eigenvalues of F on the space $H_c^1(U, \mathscr{V})$ have *absolute values* $\leqslant q^{(a+2)/2}$. Indeed, it follows easily from the lemma that the infinite product

$$Z_1(t) = \prod_{x \in |U_0|} 1/\det(1 - t^{\deg(x)}; \mathscr{V}_x)$$

is absolutely convergent for $|t| < q^{-(a+2)/2}$, hence it is a function without zeros or poles in that disk. But if F had an eigenvalue α in $H_c^1(U, \mathscr{V})$, with absolute value $> q^{(a+2)/2}$, α^{-1} would be a zero of $P_1^1(t)$ and not of $P_1^2(t)$; by (84) it would be a zero of $Z_1(t)$, which is absurd.

138. These preliminary results being proved, we now return to the end of part A of the proof. The space E of evanescent cocycles in $H^{n-1}(X_u, \mathbf{Q}_l)$ is the fiber of a locally constant sheaf $\mathscr{E}_0$ on $U_0 = D_0 - S_0$, which is a subsheaf of $R^{n-1} f_{0*} \mathbf{Q}_l$; there is a perfect pairing

$$\mathscr{E}_0/(\mathscr{E}_0 \cap \mathscr{E}_0^\perp) \times \mathscr{E}_0/(\mathscr{E}_0 \cap \mathscr{E}_0^\perp) \to \mathbf{Q}_l(1-n).$$

One filters $R^{n-1} f_{0*} \mathbf{Q}_l$ by the subsheafs $j_* \mathscr{E}_0$ and $j_*(\mathscr{E}_0 \cap \mathscr{E}_0^\perp)$ ($j : U_0 \to D_0$ is the natural injection). Using the exact sequence of cohomology, the exact sequence (52) of (IX, 100), and the fact that $H^1(D, \mathscr{V}) = 0$ if $\mathscr{V}$ is a constant sheaf, one is finally reduced to applying the fundamental lemma of (IX, 134) to $\mathscr{V}_0 = \mathscr{E}_0/(\mathscr{E}_0 \cap \mathscr{E}_0^\perp)$.

139. Assumption (ii) of this fundamental lemma is satisfied, due to the Kazhdan-Margulis theorem (IX, 102), but there is still to be proved that assumption (iii) of the lemma holds. This again needs some rather intricate arguments.

One first applies the Grothendieck formula (77), but this time to each single fiber X_x for $x \in |U_0|$ (instead of X_0). This results in the existence of a finite set of

l-adic units α_i $(1 \leqslant i \leqslant Q)$, β_j $(1 \leqslant j \leqslant R)$, all distinct, such that for *every* $x \in |U_0|$, the product

$$\frac{\prod_i (1 - \alpha_i^{\deg(x)} t) \, . \, \det(1 - F_x t; \mathscr{V}_x)}{\prod_j (1 - \beta_j^{\deg(x)} t)} \tag{88}$$

is a rational function of t with *coefficients in* $\mathbf{Q}$. First we must prove that the set L of points $x \in |U_0|$ such that some $\beta_j^{\deg(x)}$ is an eigenvalue of F_x acting on $\mathscr{V}_x = \mathscr{E}_x / (\mathscr{E}_x t \mathscr{E}_x^{\perp})$ is "sparse" in the infinite set $|U_0|$. This is done by invoking the Cebotarev density theorem from class field theory, which enables one to show that L has "density 0." Before doing this, care must be taken that the action by ρ_0 of $\pi_1^{\text{alg}}(U_0)$ on $\mathscr{V}_u$ is *not* a symplectic transformation as the action of $\pi_1^{\text{alg}}(U)$, but a symplectic *similitude*, which brings additional difficulties. Finally, one gets a large enough set of points $x \in |U_0|$ for which the numerator and denominator of (88) have no common factor in $\mathbf{Q}_l[t]$, and from there assumption (iii) of (IX, 134) follows by ingenious manipulations of rational functions, thus completing the proof of (73).

140. *B) The Hard Lefschetz Theorem.* Let X be a projective irreducible smooth variety of dimension n over an algebraically closed field k. If l is a prime distinct from the characteristic of k, there is still a natural map that, to an algebraic cycle of codimension r on X (IX, 55), assigns a cohomology class in $H^{2r}(X, \mathbf{Q}_l)$. In particular, let ξ be the class in $H^2(X, \mathbf{Q}_l)$ of a smooth hyperplane section of X, and let

$$L : H^{\bullet}(X, \mathbf{Q}_l) \to H^{\bullet+2}(X, \mathbf{Q}_l)$$

be the cupproduct by ξ (IX, 87). The "hard Lefschetz theorem" is the statement that for $i < n$, the map

$$L^{n-i} : H^i(X, \mathbf{Q}_l) \to H^{2n-i}(X, \mathbf{Q}_l)$$

is an *isomorphism*; that is, the analog for l-adic cohomology of the theorem stated by Lefschetz for usual cohomology $H^{\bullet}(X, \mathbf{C})$ of complex smooth varieties, and completely proved by Hodge for Kähler compact manifolds (VII, 8).

141. The theorem was one of the "standard conjectures" (IX, 128); however, it has been *deduced* by Deligne from his proof of the Weil conjectures. We can mention only some of the main points of the very long and intricate proof, in which, even more than in the proof of the Weil conjectures, the most unexpected tools from all parts of mathematics are cleverly put to use.

Deligne first restricts himself to the case in which $k = \overline{\mathbf{F}}_q$ and $X = X_0 \otimes_{\mathbf{F}_q} \overline{\mathbf{F}}_q$ where X_0 is as in (IX, 124). Using induction on the dimension n of X, and the weak Lefschetz theorem applied to a smooth hyperplane section of X, one is reduced to the case $i = 1$ of the theorem. Then, by the monodromy technique of (IX, 132), it is seen that the theorem is equivalent to the relation $E \cap E^{\perp} = 0$ for the space of evanescent cocycles E (notation of (IX, 138)). (In fact, this is substantially the way Lefschetz had stated the theorem.) The Kazhdan-Margulis theorem (IX, 102) shows that *if* $E \cap E^{\perp} = 0$, the linear representation $\rho = \rho_0 | \pi_1^{\text{alg}}(U): \pi_1^{\text{alg}}(U) \to GL(E)$ is absolutely irreducible. But *conversely*, if ρ is *semi-simple*, it follows easily that $E \cap E^{\perp} = 0$.

Now the proof of the Weil conjecture (69) used essentially properties of the representation ρ; Deligne's idea is to prove the hard Lefschetz theorem by *reversing* the process, showing that the Weil conjecture implies that ρ is semi-simple. There are in $\pi_1^{\text{alg}}(U_0)$ elements f_x (for each $x \in |U_0|$) that are well defined up to conjugation and that map by ρ_0 to the restriction of $F^{\deg(x)}$ to $\mathcal{O}_x$. Suppose one has, as in the fundamental lemma of (IX, 134), a locally constant $\mathbf{Q}_l$-constructible sheaf $\mathscr{V}_0$ on U_0, and the fundamental group $\pi_1^{\text{alg}}(U_0)$ operates on each fiber $\mathscr{V}_x$ of $\mathscr{V}_0$ for $x \in |U_0|$, hence $\rho_0(f_x)$ is an endomorphism of $\mathscr{V}_x$. One says ρ_0 is *pure of weight* $a \in \mathbf{Z}$ if for *any* $x \in |U_0|$ and *any* eigenvalue β of $\rho_0(f_x)$, one has, for an isomorphism of $\overline{\mathbf{Q}}_l$ onto $\mathbf{C}$,

(89) $$|\beta| = q_x^{a/2}.$$

The Weil conjecture (69) implies that $\rho_0: \pi_1^{\text{alg}}(U_0) \to GL(E)$ is *pure of weight* $n-1$. The hard Lefschetz theorem will therefore be proved if it is shown that, *in general*, if a representation $\rho_0: \pi_1^{\text{alg}}(U_0) \to GL(\mathscr{V}_u)$ is *pure of weight a*, then the restriction ρ of ρ_0 to $\pi_1^{\text{alg}}(U)$ is *semi-simple*.

142. In the rational function

(90) $$L(\rho_0, t) = \det(1 - tF; H_c^1(U, \mathscr{V}))/\det(1 - tF; H_c^2(U, \mathscr{V})),$$

one substitutes q^{-s} for t, so that (after identification of $\overline{\mathbf{Q}}_l$ and $\mathbf{C}$) one obtains a *Dirichlet series* $L_{\rho_0}(s)$. The second step in the proof is to show that if ρ_0 is of weight a, and *in addition* the following result is true:

(HV) $$L_{\rho_0}(s) \neq 0 \quad \textit{for} \quad \operatorname{Re} s = 1 + a/2,$$

then ρ is semi-simple.

143. One considers the largest quotient V_ρ of the ρ_0-module $V = \mathscr{V}_u$, on which the representation ρ acts trivially; V_ρ is isomorphic to $H_c^2(U, \mathscr{V})$, and one first proves that the dimension of V_ρ is equal to the *number of poles* of $L(\rho_0, t)$. This follows from the fact that there is no cancellation between the numerator and denominator of (90), and this last property turns out to be a consequence of the fact that ρ_0 is of weight a and of (HV), using also the "Riemann hypothesis for curves," first when ρ_0 is irreducible, and then by induction on the length of the ρ_0-module V.

Once this is done, the semi-simplicity of ρ is shown by comparing the representation ρ_0 in V with the *semi-simple* representation ρ_0^{ss} in the graded ρ_0-module V^{ss}, the direct sum of the quotients of a Jordan-Hölder sequence for the ρ_0-module V. One has $L(\rho_0, t) = L(\rho_0^{ss}, t)$. The interpretation of dim V_ρ by the poles of $L(\rho_0, t)$ therefore shows that V_ρ has the same dimension as the similar module V_ρ^{ss}. This is also true of the ρ-module V^ρ of invariants of V under ρ and of the similar ρ-module $(V^{ss})^\rho$. One ends this part of the proof by showing that $V \mapsto V^\rho$ is an exact functor on the category of ρ_0-modules of given weight, and using this, one proves that any exact sequence $0 \to V' \to V \to V'' \to 0$ of ρ_0-modules of same weight splits, observing that Hom (V'', V) and Hom (V'', V'') have weight 0, and that Hom $(V'', V)^{\rho} \to$ Hom $(V'', V'')^{\rho}$ is surjective.

144. The last step of the proof consists in showing that *if* ρ_0 is of weight a, then (HV) *is true*. It is too long and intricate to be summarized here. We simply note that it draws on the theory of affine group schemes of Grothendieck, on the theory of complex semi-simple Lie groups, and finally on the analytic device by which Hadamard and de la Vallée-Poussin proved that the Riemann zêta function has no zeros on the line Re $s = 1$.

145. Once Deligne had proved the hard Lefschetz theorem over the field $\bar{\mathbf{F}}_q$, he was also able to extend it to an *arbitrary* algebraically closed field k. The standard techniques of the theory of schemes allow one to construct, for a projective smooth connected variety X over k, a noetherian scheme S for which a prime number l is invertible in all residual fields $\kappa(s)$, $s \in S$, and a projective smooth map $f: Y \to S$, with a fiber over the geometric generic point $\bar{\eta}$ being the given X, and some special fiber Y_s being defined over $\kappa(s) = \mathbf{F}_q$. The properties of "base change" in l-adic cohomology then show that the maps $L_Y^{n-i}: R^i f_*(\mathbf{Q}_l) \to R^{2n-i} f_*(\mathbf{Q}_l)$ are isomorphisms if and only if they induce isomorphisms $H^i(Y_{\bar{s}}) \to H^{2n-i}(Y_{\bar{s}})$ for one $s \in S$.

146. *C) Abelian Varieties over Finite Fields.* Let A be an abelian variety defined over a finite field $\mathbf{F}_q$. It follows from Deligne's theorem (but was already proved by Weil in this case) that the eigenvalues of the Frobenius morphism acting on $H^1(A, \mathbf{Q}_l)$ are algebraic numbers whose conjugates all have absolute values $q^{1/2}$. It is easy to see that two *isogenous* abelian varieties (VII, 56) over $\mathbf{F}_q$ have the *same* zêta function. Conversely, Tate has proved that the eigenvalues of F acting on $H^1(A, \mathbf{Q}_l)$, counted with their multiplicities, determine the isogeny class of A. Furthermore, Honda has shown that for *any* algebraic number α such that all its conjugates have absolute value $q^{1/2}$, there is an abelian variety A over $\mathbf{F}_q$ having α as one of the eigenvalues of the Frobenius morphism acting on $H^1(A, \mathbf{Q}_l)$.

Tate has also proved that if A and B are two abelian varieties defined over a finite field $\mathbf{F}_q$, the natural map of (IX, 111, formula (60)) is *bijective*.

147. *D) Curves and Abelian Varieties Defined over a Number Field.* Among the oldest problems of mathematics is the search for solutions in *rational numbers* of algebraic equations with *rational* coefficients, since numerous problems of this kind were already treated by Diophantus.

For equations in *two* variables, the modern formulation of the problem is the following one: given a projective smooth irreducible curve C defined over a *number field* K, what can one say of the set $C(K')$ of points of C "with values in K′" (VII, 25 and VIII, 43), when K′ is a *finite* extension of K? It is now known that the answer depends on the *genus* of C:

1) If $g = 0$, either $C(K') = \emptyset$ or $C(K')$ is isomorphic with the projective line $\mathbf{P}_1(K')$; this was already known to M. Noether (VII, 34).

2) If $g = 1$, either $C(K') = \varnothing$, or the image of $C(K')$ in the jacobian $J(C)$ (VII, 57) has the form $z_0 + G$, where G is a finitely generated commutative subgroup of $J(C)$. This was proved by Mordell in 1922 (VII, 35).

3) If $g \geqslant 2$, $C(K')$ is *finite*. This was conjectured by Mordell, but only proved in 1983 by G. Faltings.

148. The results for $g \geqslant 1$ are *noneffective*. For $g = 1$, no general result is known on the group G. For $g \geqslant 2$, nothing is known in general on the number of elements of $C(K')$, nor, if C is embedded in $\mathbf{P}_2$, how big the coordinates of these elements may be. For instance, if $m \geqslant 4$, the number of rational solutions of $ax^m + by^m = c$, where a, b, c are integers $\neq 0$, is finite by Faltings's theorem, but nothing is known in general on this number.

149. The Mordell theorem for jacobians was generalized by A. Weil to all abelian varieties, and it is by using this theorem that Siegel had been able to prove a part of the Faltings's result, namely that for affine curves in $\mathbf{P}_2$ of genus $\geqslant 1$, the number of points of $C(K')$ whose coordinates are *integers* in K′ is finite (VII, 35).

150. It is also by a deep study of abelian varieties defined over a number field K that Faltings has proved his theorem. Such a variety A has a *Néron model*, which is a well-determined smooth group scheme (VIII, 44) X over $\mathrm{Spec}(\mathfrak{o})$, where $\mathfrak{o}$ is the ring of integers of the number field K. It is such that the fiber of X at the generic point $\mathrm{Spec}(K)$ of $\mathrm{Spec}(\mathfrak{o})$ is isomorphic to A. For any point $s \in \mathrm{Spec}(\mathfrak{o})$ (i.e. a prime ideal of $\mathfrak{o}$), the fiber X_s is then a group variety over the field $\kappa(s)$; one says that X (or A) has *good reduction at s* if X_s is an abelian variety.

151. The novelty in Faltings's approach is the notion of *height* $h(A)$ of an abelian variety A defined over a number field K. It is a real number that cannot be described here in detail, and that is first constructed "locally" at each place of K, using the Néron model at finite places, and the embeddings of K in $\mathbf{C}$ at places at infinity.

A fundamental technical result is then that, for a given triple (g, d, h), there is only a *finite* number of isomorphism classes of *polarized* abelian varieties of dimension g, degree of polarization d (IX, 113), and height $h(A) \leqslant h$. The proof is long and intricate, drawing on many tools from analysis and algebraic geometry; its main idea is to compare the height $h(A)$ to the height (in a sense that is classical in number theory) of the point corresponding to A in the *scheme of moduli* $A_{g,d,m}$ (IX, 115).

152. Another important technical result concerns a sequence (A/K_n) of quotients of A defined as follows. On the Tate module $T_l(A)$ (VII, 61), Weil has defined an alternating bilinear nondegenerate form

$$T_l(A) \times T_l(A) \to \mathbf{Z}_l(1). \tag{91}$$

On the other hand the Galois group $G = \mathrm{Gal}(\bar{K}/K)$ operates on $T_l(A) \otimes_{\mathbf{Z}_l} \mathbf{Q}_l$. Let W be a G-stable maximal isotropic subspace of $T_l(A) \otimes_{\mathbf{Z}_l} \mathbf{Q}_l$, and let K_n be the image of $T_l(A) \cap W$ in the kernel A_{l^n} of the endomorphism $z \mapsto l^n . z$ of A, which is also equal to $T_l(A)/l^n T_l(A)$. Then Faltings proves that, for large enough n, the height $h(A/K_n)$ is *independent of* n; this time, one has to use results of Raynaud on commutative group schemes, and the Weil theorem on the eigenvalues of the Frobenius morphism on $H^1(A, \mathbf{Q}_l)$ (IX, 146).

153. Joined to the finiteness result of (IX, 151), this enables Faltings to prove:

1) Over a number field, the natural map of (IX, 111, formula (60)) is *bijective*, a conjecture of Tate.

2) Over a number field, an isogeny class of abelian varieties only contains a *finite* number of isomorphism classes.

3) Let S be a finite set of places of a number field K; the number of isomorphism classes of abelian varieties over K of dimension g (resp. of polarized abelian varieties over K, of dimension g and degree of polarization d) having *good reduction* (IX, 150) outside S, is *finite*, a conjecture of Shafarevich.

The Torelli theorem, under the form (IX, 114) shows that one can deduce from the Shafarevich conjecture that the number of isomorphism classes of curves of given genus over K, such that their jacobian has good reduction outside S, is also *finite*.

154. Let C be a curve of genus $g \geqslant 2$ over K. A construction due to Kodaira and Parshin provides a finite extension K′ of K, a finite set S′ of places of K′, and a set U′ defined as the set of isomorphism classes of curves of genus g' given by $2g' - 2^{2g}(4g - 3)$, defined over K′, and whose jacobian has good reduction outside S′. The construction also provides a map $u : C(K) \to U'$, such that the inverse image by u of any element of U′ is *finite*. As U′ is itself finite by (IX, 153), one concludes that C(K) is finite, thus proving Mordell's conjecture.

9. Additional topics

155. For the following research areas, only references are given in the Bibliography:

1) Weierstrass points on curves and their generalizations.
2) Real algebraic curves.
3) Geometry of webs, and its relation with the Castelnuovo curves.
4) Relations between moduli of elliptic curves and p-adic L-functions.
5) Compactification of moduli spaces.
6) Local deformations of algebraic varieties.
7) Local cohomology and cohomology of completions of schemes.
8) Brauer groups and cohomology.
9) Motives.
10) Intersection cohomology.
11) Étale homotopy.
12) Prym varieties.
13) Algebraic structures on compact quotients of C^∞ manifolds.

Annotated Bibliography

II. For a survey of the development of Greek mathematics, see:

[1] T. HEATH, *A history of Greek mathematics*, 2 vol., Oxford (Clarendon Press), 1921.

A controversial question concerns the interpretation that can be given to certain algebraic methods of Diophantus; although no geometric notion is mentioned here, the procedures giving a rational parametric representation of simple unicursal curves are seen in filigree. For Diophantus's text, see:

[2] T. HEATH, *Diophantus of Alexandria*, 2nd ed., New York (Dover), 1964.

III. The basic work for the history of the development of algebraic geometry from 1630 to the end of the nineteenth century is the *Bericht* of Brill and M. Noether:

[3] A. BRILL and M. NOETHER, Die Entwicklung der Theorie der algebraischen Functionen in älterer und neuerer Zeit, *Jahresber. der Deutsche Math. Verein.*, t. III (1892–1893), p. 111–566.

III,6. The history of "elimination" is recounted in:

[4] E. NETTO, Rationale Funktionen mehrerer Veränderlichen, *Enzykl. der math. Wiss.*, I B 1 *b*), p. 255–282 (1894).

III,9 and 10. For more details on plane algebraic curves, consult:

[5] G. KOHN and G. LORIA, Spezielle ebene algebraische Kurven, *Enzykl. der Math. Wiss.* III C 5, *a)* and *b)*, p. 457–634 (1908–1914).

[6] G. LORIA, *Spezielle algebraische und transzendente ebene Kurven:* I. *Die algebraischen Kurven* (2te Aufl.), Leipzig-Berlin (Teubner), 1910.

IV. In addition to [3], for this period, one can consult with profit:

[7] G. FANO, Gegensatz von synthetischer und analytischer Geometrie in seiner historischen Entwicklung im XIX. Jahrhundert, *Enzykl. der math. Wiss.*, III AB 4 *a)*, p. 221–288 (1907).

[8] E. KÖTTER, Die Entwicklung der synthetischen Geometrie von Monge bis auf Staudt (1847), *Jahresber. der Deutsche Math. Verein.*, t. V (1901), p. 1–476.

IV,2. See:

[9] G. DESARGUES, *Œuvres*, t. I (éd. POUDRA), Paris, 1864.

IV,5. For a detailed account, see [8] and:

[10] F. DINGELDEY, Kegelschnitte und Kegelschnittsysteme, *Enzykl. der math. Wiss.*, III C 1, p. 1–160 (1903).

[11] O. STAUDE, Flächen 2. Ordnung und ihre Systeme und Durchdringungskurven, *Enzykl. der math. Wiss.*, III C 2, p. 161–256 (1904).

[12] P. WOOD, The twisted cubic, *Cambridge tracts* no. 14, Cambridge University Press, 1913.

IV,6. An excellent survey of the whole of "classical" algebraic geometry, giving maximum information in the smallest volume, is:

[13] J. G. SEMPLE and L. ROTH, *Introduction to algebraic geometry*, Oxford (Clarendon Press), 1949.

For more details, see [5], [6], [8] as well as:

[14] K. Rohn and L. Berzolari, Algebraische Raumkurven und abwickelbare Flächen, *Enzykl. der math. Wiss.*, III C 9, p. 1229–1436 (1926).

[15] W. F. Meyer, Spezielle algebraische Flächen, *Enzykl. der math. Wiss.*, III C 10, p. 1437–1779 (1930).

[16] G. Loria, *Curve sghembe speziali algebriche e transcendenti*, 2 vol., Bologna (Zanichelli), 1925.

[17] R. Hudson, *Kummer's quartic surface*, Cambridge University Press, 1905.

[18] C. Jessop, *Quartic surfaces with singular points*, Cambridge University Press, 1916.

[19] A. Henderson, The 27 lines upon the cubic surface, *Cambridge tracts* no. 13, 1911.

For a generalization of the theory of lines lying on a cubic surface, see:

[20] B. L. van der Waerden, Zur algebraischen Geometrie II: Die geraden Linien auf den Hyperflächen des P_n, *Math. Ann.*, t. CVIII (1933), p. 253–259.

IV,7. For a modern account and a generalization of the Plücker formulas, see:

[21] W. Pohl, Differential geometry of higher order, *Topology*, 1 (1962), p. 169–211.

IV,9. For the history of n-dimensional projective geometry, see:

[22] C. Segre, Mehrdimensionale Räume, *Enzykl. der math. Wiss.*, III C 7, p. 769–972 (1912).

For the "geometry of lines":

[23] K. Zindler, Algebraische Liniengeometrie, *Enzykl. der math. Wiss.*, III C 8, p. 973–1228 (1921).

[24] C. Jessop, *A treatise on the line complex*, Cambridge University Press, 1903.

The most complete study of the grassmannians is found in vol. I and II of:

[25] W. Hodge and D. Pedoe, *Methods of algebraic geometry*, 3 vol., Cambridge University Press, 1953–1954.

For the evolution of the idea of "coordinates" in the nineteenth century, one can consult:

[26] E. Müller, Die verschiedenen Koordinatensysteme, *Enzykl. der math. Wiss.*, III AB 7, p. 596–770 (1910).

The connection between a space curve and the complex of lines meeting it is introduced in:

[27] A. Cayley, On a new analytical representation of curves in space (*Coll. Papers*, t. IV, p. 446–455 and 490–494).

IV,10. For an account of the theory of invariants in the nineteenth century, see:

[28] W. F. Meyer, Bericht über den gegenwärtigen Stand der Invariantentheorie, *Jahresber. der Deutsche math. Verein.*, t. I (1890–1891), p. 79–292.

More modern presentations are given in:

[29] H. Weyl, *The classical groups*, Princeton Univ. Press, 1939.

[30] J. Dieudonné and J. Carrell, *Invariant theory, old and new*, New York and London (Academic Press), 1971.

IV,11. An account of the whole of the work on transformations and correspondences in algebraic geometry in the nineteenth century is furnished by:

[31] L. Berzolari, Algebraische Transformationen und Korrespondenzen, *Enzykl. der math. Wiss.*, III C 11, p. 1781–2218 (1932).

IV,13. For an account of Chasles's ideas, with numerous examples, see:

[32] T. LEMOYNE, *Les lieux géométriques...*, Paris (Vuibert), 1923.

Chasles's theory is presented in a rigorous manner in

[33] B. L. van der WAERDEN, Zur algebraischen Geometrie XV: Lösung des Charakteristikenproblems für Kegelschnitte, *Math. Ann.*, t. CXV (1938), p. 645–655.

V. In addition to [3], classical accounts of Riemann's theory are found in:

[34] B. RIEMANN, *Gesammelte mathematische Werke*, 2d ed., Leipzig (Teubner), 1892.

[35] E. PICARD, *Traité d'Analyse*, vol. II, 3d ed., Paris (Gauthier-Villars), 1925.

[36] P. APPELL and E. GOURSAT, *Théorie des fonctions algébriques et de leurs intégrales*, vol. I (2d ed.), Paris (Gauthier-Villars), 1929.

V,3 and 4. See:

[37] N. H. ABEL, *Œuvres*, 2 vol., ed. SYLOW et LIE, Christiania, 1881.

V,6. See:

[38] E. GALOIS, *Ecrits et mémoires mathématiques*, ed. R. BOURGNE et J.-P. AZRA, Paris (Gauthier-Villars), 1962, p. 181.

V,8. The classical work of H. Weyl is:

[39] H. WEYL, *Die Idee der Riemannschen Fläche*, Leipzig (Teubner), 1913 (3. Auflage, Leipzig (Teubner), 1955).

For a modern account, see the excellent book of S. Lang:

[40] S. LANG, *Introduction to algebraic and abelian functions*, Reading (Addison-Wesley), 1972.

VI,2. See [3] and:

[41] L. KRONECKER, *Werke*, t. II, p. 237–387, Leipzig (Teubner), 1897.

[42] E. LASKER, Zur Theorie der Moduln und Ideale, *Math. Ann.*, t. LX (1905), p. 20–116.

[43] F. S. MACAULAY, On the resolution of a given modular system into primary systems including some properties of Hilbert numbers, *Math. Ann.*, t. LXXIV (1913), p. 66–121.

[44] G. LANDSBERG, Algebraische Gebilde; arithmetische Theorie algebraischer Grössen, *Enzykl. der math. Wiss.*, I B 1*c)*, I C 5, p. 283–319 (1904).

VI, 3. See:

[45] R. DEDEKIND, *Gesammelte mathematische Werke*, t. I, p. 238–350, Braunschweig (Vieweg), 1932.

The Dedekind-Weber memoir is developed in:

[46] K. HENSEL and G. LANDSBERG, *Theorie der algebraischen Funktionen einer Variabeln und ihre Anwendung auf algebraische Kurven und abelsche Integrale*, Leipzig (Teubner), 1902.

For modern accounts, see [40] as well as:

[47] C. CHEVALLEY, *Introduction to the theory of algebraic functions of one variable*, A.M.S. Math. Surveys, VI, New York, 1951.

[48] J.-P. SERRE, *Groupes algébriques et corps de classes*, Paris (Hermann), 1959.

[49] W. FULTON, *Algebraic curves*, New York-Amsterdam (Benjamin), 1969.

VI,16. For classical accounts of the theory of algebraic curves, see [3], [13] as well as:

[50] L. BERZOLARI, Allgemeine Theorie der höheren ebenen algebraischen Kurven, *Enzykl. der math. Wiss.*, III C 4, p. 313–455 (1906).

[51] J. COOLIDGE, *A treatise on algebraic plane curves*, New York (Dover).

[52] L. Godeaux, *Introduction à la géométrie supérieure*, (2d ed.), Liège (Ed. Sciences et Lettres), 1946.

[53] L. Godeaux, *Géométrie algébrique*, 2 vol., Paris (Masson), 1948.

VI,19. See:

[54] L. Kronecker, *Werke*, t. II, p. 193–236, Leipzig (Teubner), 1897.

VI,24. See [13].

VI,27. For the particular cases of the theory of "moduli" studied in the nineteenth century, see [3].

VI,30 The classical work is:

[55] E. Picard and G. Simart, *Théorie des fonctions algébriques de deux variables indépendantes*, 2 vol., Paris (Gauthier-Villars), 1897–1906.

The transcendental theory of algebraic surfaces as well as that of the linear systems of the Italian school are described in great detail in:

[56] O. Zariski, *Algebraic surfaces*, Erg. der Math., Bd. III, Heft 5, Berlin (Springer), 1935; 2d. ed. Berlin-Heidelberg-New York (Springer), 1970.

The second edition contains important appendices, written by S. Abhyankar, J. Lipman and D. Mumford, interpreting in modern terms the results cited by Zariski and indicating how they have been completed or developed up to 1970.

VI,31. See [56], Chap. I, as well as:

[57] O. Zariski, Local uniformization on algebraic varieties, *Ann. of Math.*, 41 (1940), p. 852–896.

VI,32. See:

[58] R. Walker, Reduction of the singularities of an algebraic surface, *Ann. of Math.*, 36 (1935), p. 336–365.

[59] O. Zariski, Simplified proof for the resolution of the singularities of a surface, *Ann. of Math.*, 43 (1942), p. 583–593.

VI,33. See [13].

VI,36. See:

[60] S. Lefschetz, *Topology*, New York (A.M.S. Coll. Publ. no. 12), 1930.

[61] S. Lefschetz, *Selected Papers*, New York (Chelsea), 1971.

VI,37. A proof of the triangulability of algebraic varieties is found in:

[62] B. L. van der Waerden, *Einführung in die algebraische Geometrie*, Berlin (Springer), 1939.

Lefschetz's results on the topology of algebraic varieties are described in [61] and:

[63] S. Lefschetz, *L'Analysis situs et la Géométrie algébrique*, Paris (Gauthier-Villars), 1924.

The proofs, sometimes sketchy, of this work are developed in detail in:

[64] A. Wallace, *Homology theory on algebraic varieties*, London-New York-Paris-Los Angeles (Pergamon), 1958.

See also [56], Chap. VI.

VI,39. See [13], [56], [61], [63], as well as:

[65] L. Berzolari, Algebraische Transformationen und Korrespondenzen, *Enzykl. der math. Wiss.*, III C 11, p. 1781–2218 (1932).

VI,43. For a classical account of the results of the Italian school on the theory of algebraic surfaces, see above all [56], as well as [13], [53], [55], and:

[66] G. Castelnuovo and F. Enriques, Grundeigenschaften der algebraischen Flächen,

Enzykl. der math. Wiss., III C 6 *a)*, p. 635–673 (1908).

[67] G. Castelnuovo and F. Enriques, Die algebraischen Flächen vom Gesichtspunkte der birationalen Transformationen aus, *Enzykl. der math. Wiss.*, III C 6 *b)*, p. 674–767 (1914).

[68] F. Severi, *Vorlesungen über algebraische Geometrie*, Leipzig (Teubner), 1921.

For the corresponding results for projective algebraic varieties of dimension $\geqslant 3$, see:

[69] L. Roth, *Algebraic threefolds*, Erg. der math., Neue Folge, Heft 6, Berlin-Heidelberg-New York (Springer), 1955.

VI,47. See:

[70] F. Enriques, Sui sistemi continui di curve appartenenti ad una superficie algèbrica, *Comm. Math. Helv.*, 15 (1943), p. 227–237.

[71] F. Severi, Intorno ai sistemi continui di curve sopra una superficie algèbrica, *Comm. Math. Helv.*, 15 (1943), p. 238–248.

VI,50. See [13], [53], as well as:

[72] L. Godeaux, *Les transformations birationnelles du plan* (2d ed.), Mém. des Sci. math., no. 122, Paris (Gauthier-Villars), 1953.

[73] L. Godeaux, *Les transformations birationnelles de l'espace*, Mém. des Sci. math., no. 67, Paris (Gauthier-Villars), 1934.

[74] H. Hudson, *Cremona transformations in plane and space*, Cambridge University Press, 1927.

VI,53. For a modern study of Del Pezzo surfaces (over a perfect field of arbitrary characteristic), see:

[75] Ju. Manin, Rational surfaces over perfect fields (en russe, avec résumé anglais), *Publ. math. I.H.E.S.*, no. 30 (1966), p. 55–113.

VI,57. For a modern study of Cremona transformations, see:

[76] M. Demazure, Sous-groupes algébriques de rang maximum du groupe de Cremona, *Ann. Ec. Norm. Sup.*, (4), 3 (1970), p. 507–588.

[77] A. Hirschowitz, Le groupe de Cremona d'après Demazure, *Sém. Bourbaki*, Exposé 413, Lecture Notes in math., no. 317, p. 261–276, Berlin-Heidelberg-New York (Springer), 1973.

VI,58. See [73], and for involutions on a surface:

[78] L. Godeaux, *Théorie des involutions cycliques appartenant à une surface algébrique et applications*, Roma (Cremonese), 1963.

VII,3. See:

[79] G. de Rham, *Variétés différentiables*, Paris (Hermann), 1955.

VII,4. See [79], as well as:

[80] W. Hodge, *The theory and applications of harmonic integrals*, 2d. ed., Cambridge University Press, 1952.

[81] A. Weil, *Variétés kählériennes*, Paris (Hermann), 1958.

[82] M. Baldassari, *Algebraic varieties*, Erg. der Math., Neue Folge, Heft 12, Berlin-Göttingen-Heidelberg (Springer), 1956.

VII,11. The first book (chronologically) treating projective algebraic geometry over an arbitrary field is [62]. Written in the same spirit are [25] and:

[83] S. Lefschetz, *Algebraic geometry*, Princeton University Press, 1953.

[84] O. Zariski, *Introduction to the theory of algebraic surfaces*, Lect. Notes in Math., no. 83, Berlin-Heidelberg-New York (Springer), 1969;

for algebraic curves, [47], [49] and:

[85] R. WALKER, *Algebraic curves*, Princeton University Press, 1950;

and for algebraic surfaces:

[86] H. JUNG, *Einführung in die Theorie der algebraischen Funktionen zweier Veränderlicher*, Berlin (Akad. Verlag), 1951.

It should be noted that in [85] and [86] the base field is restricted to be of characteristic 0.

VII,12. The memoir of F. K. Schmidt is:

[87] F. K. SCHMIDT, Analytische Zahlentheorie in Körpern der Charakteristik p, *Math. Zeitschr.*, 33 (1931), p. 1–32.

VII,15. See [41], [42] and [43], as well as:

[88] E. NOETHER, Idealtheorie in Ringbereichen, *Math. Ann.*, 83 (1921), p. 24–66.

VII,16. See [3] and:

[89] D. HILBERT, *Gesammelte Abhandlungen*, vol. II, Berlin (Springer), 1933.

VII,21. See:

[90] H. ZEUTHEN, Abzählende Methoden, *Enzykl. der math. Wiss.*, III C 3, p. 257–312 (1905).

[91] H. SCHUBERT, *Kalkül der abzählende Geometrie*, Leipzig (Teubner), 1879.

Numerous examples of Schubert calculus are developed in [13]. For a modern account of questions relating to this calculus, see [62], [95] and:

[92] S. KLEIMAN and D. LAKSOV, Schubert Calculus, *Amer. math. Monthly*, 79 (1972), p. 1061–1082.

VII,24. See [62] and:

[93] F. SEVERI, Sul principio della conservazione del numero, *Rend. Circ. Mat. Palermo*, 33 (1912), p. 313–327.

VII,25. See [62], [25], and, for Weil's point of view, [82] as well as:

[94] A. WEIL, *Foundations of algebraic geometry*, New York (A.M.S. Coll. Publ. no. 29), 1946.

[95] P. SAMUEL, *Méthodes d'algèbre abstraite en géométrie algébrique*, Erg. der Math., Neue Folge, Heft 4, Berlin-Heidelberg-New York (Springer), 1955.

[96] S. LANG, *Introduction to algebraic geometry*, New York (Interscience), 1958.

VII,29. For the definitions of Chevalley and Samuel, see [95] and:

[97] C. CHEVALLEY, Intersections of algebraic and algebroid varieties, *Trans. Amer. Math. Soc.*, 57 (1945), p. 1–85.

[98] P. SAMUEL, La notion de multiplicité en algèbre et en géométrie algébrique, *J. de Math.*, (9), 30 (1951), p. 159–274.

VII,30. Gröbner's example is given in:

[99] W. GRÖBNER, *Moderne algebraische Geometrie*, Wien-Innsbrück (Springer), 1949, p. 180.

For Serre's definition, see:

[100] J.-P. SERRE, *Algèbre locale. Multiplicités*, Lect. Notes in Math., no. 11, Berlin-Heidelberg-New York (Springer), 1965.

VII,31. See [62], [82], [95] as well as:

[101] W. L. CHOW and B. L. van der WAERDEN, Zur algebraischen Geometrie, IX, *Math. Ann.*, 113 (1937), p. 692–704.

VII,32. See [95] as well as:

[102] Séminaire C. CHEVALLEY, 2d année: *Anneaux de Chow et applications*, Paris (Secr. math., 11, rue P.-Curie), 1958.

VII,34. See:

[103] A. HARNACK, Ueber die Vieltheiligkeit der ebenen algebraischen Curven, *Math. Ann.*, 10 (1876), p. 189–198.

[104] D. HILBERT, *Gesammelte Abhandlungen*, vol. II, Berlin (Springer), 1933.

[105] I. PETROWSKY, On the topology of real plane algebraic curves, *Ann. of Math.*, 39 (1938), p. 189–209.

VII,35. See:

[106] H. POINCARÉ, *Œuvres*, t. V, p. 438–548, Paris (Gauthier-Villars), 1950.

[107] S. LANG, *Diophantine Geometry*, New York-London (Interscience), 1962.

VII,44. See [59], [57], and, for varieties of dimension 3:

[108] O. ZARISKI, Reduction of the singularities of an algebraic 3-dimensional variety, *Ann. of Math.*, 45 (1944), p. 472–542.

VII,47. See:

[109] E. ARTIN, *Collected Papers*, p. 1–94, Reading (Addison-Wesley), 1965.

VII,48. See [87].

VII,49. See

[110] H. HASSE, Über die Riemannsche Vermutung in Funktionenkörpern, *C. r. du Congrès international des Mathématiciens, Oslo, 1936*, t. I, p. 189–206.

VII,51. See:

[111] A. WEIL, *Sur les courbes algébriques et les variétés qui s'en déduisent*, Paris (Hermann), 1948.

[112] A. WEIL, On some exponential sums, *Proc. Nat. Acad. Sci. U.S.A.*, 34 (1948), p. 204–207.

VII,52. See [55], [56], [63] and [82].

VII,53. See:

[113] T. MATSUSAKA, The criteria for algebraic equivalence and the torsion group, *Amer. J. of Math.*, 79 (1957), p. 53–66.

[114] A. NÉRON, La théorie de la base pour les diviseurs sur les variétés algébriques, *Coll. Géom. alg. Liège* (1952), p. 119–127.

[115] A. MATTUCK and J. TATE, On the inequality of Castelnuovo-Severi, *Abh. Math. Sem. Univ. Hamburg*, 22 (1958), p. 295–299.

[116] A. GROTHENDIECK, Sur une note de Mattuck-Tate, *J. reine und ang. Math.*, 200 (1958), p. 208–215.

VII,54. See [3], [40] as well as:

[117] A. ROBERT, *Elliptic curves*, Lect. Notes in Math., no. 326, Berlin-Heidelberg-New York (Springer), 1973.

VII,56. See [40], [81] as well as:

[118] A. WEIL, On Picard varieties, *Amer. J. of Math.*, 74 (1952), p. 865–894.

[119] P. GRIFFITHS, Some results on algebraic cycles on algebraic manifolds, *Algebraic Geometry* (Bombay Colloquium, 1968), Oxford University Press, 1969.

VII,60. See:

[120] A. WEIL, *Variétés abéliennes et courbes algébriques*, Paris (Hermann), 1948.

[121] S. LANG, *Abelian varieties*, New York (Interscience), 1959.

[122] D. MUMFORD, *Abelian varieties*, Oxford University Press, 1970.

This last work also treats the "transcendental" theory of abelian varieties over **C**.

VII,62. See [82] as well as:

[123] A. WEIL, On the projective embedding of Abelian varieties, *Algebraic geometry and topology* (Lefschetz Symposium, 1954), Princeton University Press, 1957.

VIII. J. Leray's first memoirs are the following:

[124] J. LERAY, L'anneau spectral et l'anneau filtré d'homologie d'un espace localement compact, *J. de Math.*, (9), 29 (1950), p. 1–139.

[125] J. LERAY, L'homologie d'un espace fibré dont la fibre est connexe, *J. de Math.*, (9), 29 (1950), p. 169–213.

For a clear and concise treatment of the notions of fiber space and sheaf, see first of all:

[126] F. HIRZEBRUCH, *Topological methods in algebraic geometry*, Erg. der Math., Neue Folge, Heft 9 (3d ed.), Berlin-Heidelberg-New York (Springer), 1966.

VIII,1. See [56], Chap. IV.

VIII,4. See:

[127] A. WEIL, *Fibre spaces in algebraic geometry*, Notes by A. WALLACE, mimeogr., University of Chicago, 1949.

VIII,6. See:

[128] K. KODAIRA, The theorem of Riemann-Roch on compact analytic surfaces, *Amer. J. of Math.*, 73 (1951), p. 813–875.

[129] K. KODAIRA, The theorem of Riemann-Roch for adjoint systems on three-dimensional varieties, *Ann. of Math.*, 56 (1952), p. 298–342.

For the formula (8) of Severi, see [69].

VIII,11. See [126] and [82], as well as:

[130] P. DOLBEAULT, Sur la cohomologie des variétés analytiques complexes, *C. r. Acad. Sci.*, 236 (1953), p. 175–177.

[131] J.-P. SERRE, Un théorème de dualité, *Comm. Math. Helv.*, 29 (1955), p. 9–26.

VIII,13. See [126], [82], as well as:

[132] M. EGER, Les systèmes canoniques d'une variété algébrique à plusieurs dimensions, *Ann. Ec. Norm. Sup.*, (3), 60 (1943), p. 143–172.

[133] J. A. TODD, The arithmetical invariants of algebraic loci, *Proc. London Math. Soc.*, (2), 43 (1937), p. 190–225.

[134] E. VESENTINI, Classi caratteristiche e varietà covarianti d'immersione, *Rend. Accad. Lincei*, (8), 16 (1954), p. 199–204.

VIII,14. See [126], [82], Grauert's article in [163] and:

[135] K. KODAIRA, On Kähler varieties of restricted type, *Ann. of Math.*, 60 (1954), p. 28–48.

VIII,17. See [82], [56], Appendix to Chap. IV, as well as:

[136] D. MUMFORD, *Lectures on curves on an algebraic surface*, Ann. of Math. Studies, no. 59, Princeton University Press, 1966.

VIII,19. See:

[137] J.-P. SERRE, Faisceaux algébriques cohérents (often quoted FAC), *Ann. of Math.*, 61 (1955), p. 197–278.

VIII,23. See:

[138] A. BOREL and J.-P. SERRE, Le théorème de Riemann-Roch (d'après Grothendieck), *Bull. Soc. Math. France*, 86 (1958), p. 97–136.

[139] G. WASHNITZER, Geometric syzygies, *Amer. J. of Math.*, 81 (1959), p. 171–248.

[140] Séminaire H. CARTAN-L. SCHWARTZ, 1963–1964, *Théorème d'Atiyah-Singer sur l'indice d'un opérateur différentiel elliptique*, Paris (Secr. math., 11, rue P.-Curie), 1965.

[141] R. PALAIS, *Seminar on the Atiyah-Singer index theorem*, Ann. of Math. Studies, no. 57, Princeton University Press, 1965.

VIII,24. See:

[142] J.-P. SERRE, Géométrie algébrique et géométrie analytique (often quoted GAGA), *Ann. Inst. Fourier*, 6 (1955), p. 1–42.

VIII,25. The two fundamental books on linear algebraic groups over an algebraically closed field are:

[143] Séminaire C. CHEVALLEY, 1956–1958, *Classification des groupes de Lie algébriques* (often quoted BIBLE), 2 vol., Paris (Secr. math., 11, rue P.-Curie), 1958.

[144] J. HUMPHREYS, *Linear algebraic groups*, Berlin-Heidelberg-New York (Springer), 1975.

The generalizations to nonalgebraically closed base fields and to "group schemes" are treated in:

[145] A. BOREL and J. TITS, Groupes réductifs, *Publ. math de l'I.H.E.S.*, no. 27 (1965), p. 55–148, et Compléments . . . , *ibid*, no. 41 (1972), p. 253–276.

[146] Séminaire de Géométrie algébrique du Bois-Marie 1962–1964 (SGA 3), dirigé par M. DEMAZURE and A. GROTHENDIECK: *Schémas en groupes*, 3 vol., Lect. Notes in Math., nos. 151, 152, 153, Berlin-Heidelberg-New York (Springer), 1970.

VIII,27. There are now several books giving good introductions, with complete proofs, to modern algebraic geometry, including the elementary notions and results of the theory of schemes:

[147] W. FULTON, *Intersection Theory*, Berlin-Heidelberg-New York-Tokyo (Springer), 1984.

[148] R. HARTSHORNE, *Algebraic geometry*, Berlin-Heidelberg-New York (Springer), 1977.

[149] I. SHAFAREVICH, *Basic algebraic geometry*, Berlin-Heidelberg-New York (Springer), 1974.

The theory is developed in great detail in [146] and:

[150] A. GROTHENDIECK, Eléments de Géométrie algébrique rédigés avec la collaboration de J. DIEUDONNÉ (often quoted EGA), chap. I–IV, *Publ. math de l'I.H.E.S.*, nos. 4, 8, 11, 17, 20, 24, 28, 32 (1960–1967).

[151] A. GROTHENDIECK and J. DIEUDONNÉ, *Eléments de Géométrie algébrique*, chap. I (2d ed.), Berlin-Heidelberg-New York (Springer), 1971.

[152] Séminaire de Géométrie algébrique du Bois-Marie 1960–1961 (SGA 1), dirigé par A. GROTHENDIECK: *Revêtements étales et Groupe fondamental*, Lect. Notes in Math., no. 224, Berlin-Heidelberg-New York (Springer), 1971.

[153] Séminaire de Géométrie algébrique du Bois-Marie 1962 (SGA 2), par A. GROTHENDIECK: *Cohomologie locale des faisceaux cohérents et théorèmes de Lefschetz locaux et globaux*, Amsterdam (North Holland), 1968.

[154] Séminaire de Géométrie algébrique du Bois-Marie, 1963–1964 (SGA 4), dirigé par M. ARTIN, A. GROTHENDIECK, J.-L. VERDIER: *Théorie des Topos et cohomologie étale des schémas*, 3 vol., Lecture Notes in Math., nos. 269, 270, 305, Berlin-Heidelberg-New York (Springer), 1972–1973.

[155] R. HARTSHORNE, *Residues and duality*, Lectures Notes of a seminar on the work of A. Grothendieck given at Harvard in 1963–1964, Lecture Notes in Math., no. 20, Berlin-Heidelberg-New York (Springer), 1966.

[156] Séminaire de Géométrie algébrique du Bois-Marie (SGA 4 1/2), dirigé par P. Deligne, avec la collaboration de J. Boutot, A. Grothendieck, K. Illusie et J.-L. Verdier: *Cohomologie étale*, Lecture Notes in Math., no. 569, Berlin-Heidelberg-New York (Springer), 1977.

[157] Seminaire de Géométrie algébrique du Bois Marie (SGA 5), dirigé par A. Gro-

thendieck, avec la collaboration de I. Bucur, C. Houzel, L. Illusie, J-P. Jouanolou, J-P. Serre: *Cohomologie l-adique et fonctions L*, Lecture Notes in Math, no. 589, Berlin-Heidelberg-New York (Springer), 1977.

[158] Séminaire de Géométrie algébrique du Bois-Marie (SGA 6), dirigé par P. Berthelot, A. Grothendieck and L. Illusie: *Théorie des intersections et théorème de Riemann-Roch*, Lecture Notes in Math., no. 225, Berlin-Heidelberg-New York (Springer), 1971.

[159] Séminaire de Géométrie algébrique du Bois-Marie 1967–1969 (SGA 7 I), dirigé par A. Grothendieck: *Groupes de monodromie en Géométrie algébrique*, Lecture Notes in Math., no. 288, Berlin-Heidelberg-New York (Springer), 1972.

[160] Séminaire de Géométrie algébrique du Bois-Marie (SGA 7 II), 1967–69, par P. Deligne and N. Katz: *Groupes de monodromie en Géométrie algébrique*, Lecture Notes in Math., no. 340, Berlin-Heidelberg-New York (Springer), 1973.

[161] A. Grothendieck, *Fondements de la Géométrie algébrique* (extraits du *Sém. Bourbaki* 1957–1962), Paris (Secr. math., 11, rue P.-Curie), 1962.

VIII,41. See:

[162] J.-P. Serre, Exemples de variétés projectives en caractéristique p non relevables en caractéristique 0, *Proc. Nat. Acad. Sci. U.S.A.*, 47 (1961), p. 108–109.

VIII,43. See [136] and [158].

VIII,45. See [102], exposé 1:

VIII,46. See [154], and Séminaire Bourbaki no. 256, and for a more elementary account:

[163] M. Artin, *Grothendieck Topologies*, Harvard University, mimeogr., 1962.

VIII,47. See Séminaire Bourbaki no. 363, and:

[164] D. Knutson, *Algebraic spaces*, Lect. Notes in Math., no. 203, Berlin-Heidelberg-New York (Springer), 1971.

IX. Important survey articles on algebraic geometry can be found in proceedings of Congresses and Symposia covering many parts of mathematics, such as:

[165] *Proceedings of the International Congress of mathematicians (Cambridge, 1958)*, ed. J.A. Todd, Cambridge University Press, 1960.

[166] *Proceedings of the International Congress of mathematicians (Stockholm, 1962)* Uppsala (Almqvist and Wiksells), 1963.

[167] *Actes du Congrès International des mathématiciens (Nice, 1970)*, Paris (Gauthier-Villars), 1971, vol. I and II.

[168] *Proceedings of the International Congress of mathematicians (Vancouver, 1974)*, ed. R. D. James, Canadian Mathematical Congress 1975, 2 vol.

[169] *Proceedings of the International Congress of mathematicians (Helsinki, 1978)*, ed. O. Lehto, Academia Scientiarum Fennicae, 1980, 2 vol.

[170] *Mathematical developments arising from Hilbert problems (Proceedings of Symposia in Pure mathematics*, vol. XXVIII), American mathematical Society, 1976, 2 vol.

Since 1955, many Conferences and Symposia have been devoted to algebraic geometry, totally or partially:

[171] *Algebraic geometry and topology* (A symposium in honor of S. Lefschetz) Princeton University Press, 1957.

[172] *Arithmetic algebraic geometry (Purdue, 1963)*, New York (Harper and Row), 1965.

[173] *Global Analysis (Papers in honor of K. Kodaira)*, Princeton University Press, 1969.

[174] *Algebraic geometry, Bombay Colloquium, 1968*, Oxford University Press, 1969.

[175] *Algebraic geometry, Oslo, 1970* (ed. F. Oort), Groningen (Wolters-Noordhoff), 1972.
[176] *Contributions to Analysis (Papers dedicated to L. Bers)*, New York and London (Academic Press), 1974.
[177] *Algebraic geometry, Arcata, 1974 (Proceedings of Symposia in Pure mathematics*, vol. XXIX), Providence (American mathematical Society), 1975.
[178] *Real and complex singularities, Oslo, 1976* (ed. P. Holm), (Sijthoff-Noordhoff), 1977.
[179] *Complex Analysis and Algebraic geometry (dedicated to K. Kodaira)*, Cambridge Univ. Press, 1977.
[180] *Singularités des surfaces (Séminaire de l'Ecole Polytechnique, 1976–77)*, Lecture Notes in Math., no. 777, Berlin-Heidelberg-New York (Springer).
[181] *Algebraic geometry (Tromsø, 1977)* (ed. L. Olson), Lecture Notes in Math., no. 687, Berlin-Heidelberg-New York (Springer), 1978.
[182] *Variétés analytiques compactes* (*Nice, 1977*), (ed. Y. Hervier and A. Hirschowitz), Lecture Notes in Math., no. 683, Berlin-Heidelberg-New York (Springer), 1978.
[183] *Surfaces algébriques (Séminaire d'Orsay, 1976–78)* (ed. J. Giraud, L. Illusie and M. Raynaud), Lecture Notes in Math., no. 868, Berlin-Heidelberg-New York (Springer), 1981.
[184] *Algebraic geometry (Copenhagen, 1978)* (ed. K. Lønsted), Lecture Notes in Math., no. 732, Berlin-Heidelberg-New York (Springer), 1979.
[185] *Algebraic geometry (University of Illinois at Chicago Circle, 1980)* (ed. A. Libgaber and P. Wagreich), Lecture Notes in Math., no. 862, Berlin-Heidelberg-New York (Springer), 1981.
[186] *Algebraic geometry (La Rabida, 1981)* (ed. J. M. Aroca, R. Buchweitz, M. Giusti and M. Merle), Lecture Notes in Math., no. 961, Berlin-Heidelberg-New York (Springer), 1982.

IX,3 to 8. See Séminaire Bourbaki nos. 417 and 571.

IX,9 to 11. See [181] (Gruson-Peskine), and Séminaire Bourbaki no. 592.

IX,13. See Séminaire Bourbaki nos. 168 and 277, and:

[187] Séminaire H. Cartan, 1960–61, *Familles d'espaces complexes et fondements de la Géométrie analytique*, Paris (Secr. math., 11, rue P. Curie), 1962.

IX,14. See Séminaire Bourbaki, nos. 383, 385 and 615, [175] (Mumford), and:

[188] D. Mumford, *Geometric invariant theory*, Erg. der Math., Neue Folge, Heft 34, Berlin-Heidelberg-New York (Springer), 1965.
[189] P. Deligne and D. Mumford, The irreducibility of the space of curves of given genus, *Publ. math. de l'I.H.E.S.*, no. 36 (1969), p. 75–109.

IX,15. See Séminaire Bourbaki, nos. 129, 136, 167, 216 and 389.

IX,16 to 26, see:

[190] I. Shafarevich *et al.*, Algebraic surfaces, *Proc. Steklov Inst. of Math.*, no. 75, Amer. Math. Soc., 1957.
[191] A. Beauville, Surfaces algébriques complexes, *Astérisque*, no. 54 (1978).

IX,17 and 18. See Séminaire Bourbaki, no. 146, and:

[192] I. Shafarevich, *Lectures on minimal models and birational transformations*, Tata Inst. of Fund. Research, Bombay (mimeogr.), 1966.
[193] O. Zariski, *Introduction to the problem of minimal models in the theory of algebraic surfaces*, Publ. Math. Soc. Japan, no. 4, Tokyo, 1958.

IX,19 to 26, see Séminaire Bourbaki, nos. 500 and 506, [177] (Bombieri-Husemoller) and [179].

IX,23. See Séminaire Bourbaki, nos. 609 and 611.

IX,25. See:

[194] E. Bombieri, Canonical models of surfaces of general type, *Publ. math. de l'I.H.E.S.*, no. 42 (1973), p. 171–220.

IX,27 to 32. See Séminaire Bourbaki, no. 568, and:

[195] K. Ueno, *Classification theory of algebraic varieties and compact complex spaces* (in collaboration with P. Cherenack), Lecture Notes in Math., no. 439, Berlin-Heidelberg-New York (Springer), 1975.

IX,33. See Séminaire Bourbaki, no. 402, and:

[196] R. Hartshorne, Varieties of small dimension in projective spaces, *Bull. Amer. Math. Soc.*, 80 (1974), p. 1017–1032.

IX,33. See:

[197] H. Popp, *Moduli theory and classification theory of algebraic varieties*, Lecture Notes in Math. no. 620, Berlin-Heidelberg-New York (Springer) 1977.

IX,36 to 39. See Séminaire Bourbaki, no. 530, [196] and:

[198] R. Hartshorne, Algebraic vector bundles on projective spaces: a problem list, *Topology*, 18 (1979), p. 117–118.

[199] C. Okonek, M. Schneider, H. Spindler, *Vector bundles on complex projective spaces*, Boston-Basel-Stuttgart (Birkhäuser), 1980.

IX,40. See Séminaire Bourbaki, no. 519.

IX,41. See Séminaire Bourbaki, no. 473.

IX,42. See [177] (P. Wagreich).

IX,43–44–45. See Séminaire Bourbaki, no 565, [76], [77], and:

[200] G. Kempf, F. Knudsen, D. Mumford, D. Saint-Donat, *Toroidal embeddings* I, Lecture Notes in Math., no. 339, Berlin-Heidelberg-New York (Springer), 1973.

[201] *Seminar on differential geometry* (ed. S. Yau), Ann. of Math. Studies no. 102, Princeton Univ. Press, 1982 (B. Teissier).

IX,46. See Séminaire Bourbaki, nos. 320 and 344, [177] (J. Lipman), and:

[202] H. Hironaka, Resolution of singularities of an algebraic variety over a field of characteristic zero, *Ann. of Math.* 79 (1964), p. 109–326.

[203] S. Abhyankar, *Resolution of singularities of embedded algebraic surfaces*, New York-London (Academic Press), 1966.

IX,47. See [178] (P. Orlik) and:

[204] J. Milnor, *Singular points of complex hypersurfaces* (Ann. of Math. Studies, no. 61), Princeton University Press, 1968.

IX,48 to 51. See Séminaire Bourbaki no. 250, [180] and:

[205] D. Mumford, The topology of normal singularities of an algebraic surface and a criterion for simplicity, *Publ. math. de l'I.H.E.S.*, no. 9 (1961).

[206] J. Lipman, Rational singularities with applications to algebraic surfaces and unique factorization, *Publ. math. de l'I.H.E.S.*, no. 36 (1969).

IX,52. See [177] (B. Teissier) and [180] (B. Teissier).

IX,53 to 58. See [177] (R. Hartshorne), [175] (S. Kleiman, D. Lieberman), [119] and:

[207] D. Lieberman, Higher Picard varieties, *Amer. Journ. of Math.*, 90 (1968) p. 1165–1199.

IX,53 to 60. See [177] (Hartshorne).

IX,56 and 57. See Séminaire Bourbaki no. 376, and [119].

IX,57. See [175] (Kleiman, Lieberman) and [207].

IX,59. See:

[208] H. Hironaka, Smoothing of algebraic cycles of small dimension, *Amer. Journ. of Math.*, 90 (1968), p. 1–54.

[209] S. Kleiman, Geometry of Grassmannians and applications to splitting bundles and smoothing cycles, *Publ. math. de l'I.H.E.S.*, 36 (1969), p. 281–298.

IX,60 to 62. See Séminaire Bourbaki, no. 301.

IX,63. For Nagata's example, see [192], p. 119. For Hironaka's example, see:

[210] H. Hironaka, An example of a non-kählerian deformation, *Ann. of Math.* 75 (1962), p. 190–208.

IX,64. See Séminaire Bourbaki, no. 493, and [177] (Lieberman-Mumford).

IX, 65. See:

[211] R. Hartshorne, *Ample subvarieties of algebraic varieties*, Lecture Notes in Math., no. 156, Berlin-Heidelberg-New York (Springer), 1970.

IX,66. See Séminaire Bourbaki, no. 544.

IX,68. See Séminaire Bourbaki, no. 75, and [48].

IX,69 to 71. See Séminaire Bourbaki, no. 464, [177] (Fulton) and:

[212] P. Baum, W. Fulton and R. MacPherson, Riemann-Roch for singular varieties, *Publ. math. de l'I.H.E.S.*, 45 (1975), p. 101–145.

IX,72. See:

[213] W. Fulton, Rational equivalence on singular varieties, *Publ. math. de l'I.H.E.S.*, 45 (1975), p. 147–167.

IX,74. See Séminaire Bourbaki, no. 149, and:

[214] A. Altman and S. Kleiman, *Introduction to Grothendieck Duality Theory*, Lecture Notes in Math., no. 146, Berlin-Heidelberg-New York (Springer), 1970.

IX, See Séminaire Bourbaki, no. 300, [154], [155] and [184] (Mebkhout).

IX,76 to 82. See [170] (Kleiman).

IX,77. See:

[215] C. Ehresmann, Sur la topologie de certains espaces homogènes, *Ann. of Math.*, 35 (1934), p. 396–443.

[216] G. Z. Giambelli, Sul principio della conservazione del numero, *Jahresber. der D.M.V.*, 13 (1904), p. 545–556.

[217] A. Lascoux, Puissances extérieures, déterminants et cycles de Schubert. *Bull. Soc. math. de France*, 102 (1974), p. 161–179.

[218] *Singularités d'applications différentiables (Plans-sur-Bex, 1975)*, (ed. O. Burlet and F. Ronga), Lecture Notes in Math., no. 535, Berlin-Heidelberg-New York (Springer), 1976. (A. Lascoux).

IX,81. See:

[219] E. Study, Über die Geometrie der Kegelschnitte, insbesondere deren Charakteristiken Problem, *Math. Ann.*, 26 (1886), p. 58–101.

IX,83. See [177] (Griffiths-Cornalba), Séminaire Bourbaki, no. 376 and:

[220] P. Deligne, Théorie de Hodge: II, *Publ. math. de l'I.H.E.S.*, 40 (1973), p. 5–58; III, *ibid.*, 44 (1974), p. 5–78.

IX,84. See [56], Appendix to chap. VII, and [136].

IX, 85 to 87. Up to now, the only complete proofs in the literature are to be found in [154], [156], and [157].

IX,88. See:

[221] *Dix exposés sur la cohomologie des schémas* (Grothendieck), Amsterdam-London (North Holland), 1968.

IX,89. See [221] (Grothendieck), [177] (Illusie), and:

[222] Journées de Géométrie algébrique de Rennes (1978), *Astérisque*, 63–65, 1978 (Berthelot-Messing).

[223] P. Berthelot, *Cohomologie cristalline des schémas de caractéristique* p > 0, Lecture Notes in Math., no. 407, Berlin-Heidelberg-New York (Springer), 1974.

IX,90. See [221] (Grothendieck) and:

[224] A. Grothendieck, On the De Rham cohomology of algebraic varieties, *Publ. math. de l'I.H.E.S.*, 29 (1966), p. 95–103.

IX,91. See [222] (Illusie).

IX,93 to 97. See Séminaire Bourbaki, no. 204.

IX,98. See Séminaire Bourbaki, no. 543 and:

[225] J.-P. Serre, On the fundamental group of a unirational variety, *Journ. London Math. Soc.*, 34 (1959), p. 481–484.

IX,99 to 102. Up to now, the only complete proofs in the literature are to be found in [152], [159] and [160].

IX,103 to 106. See [177] (Hartshorne) and [211].

IX,104. See:

[226] M. Kneser, Über die Darstellung algebraischen Raumkurven als Durchschnitte von Flächen, *Archiv der Math.*, 11 (1960), p. 157–158.

[227] M. P. Murthy, Complete intersections, Conf. on commutative algebra, Kingston 1975, *Queen's papers* 42 (1975), Queen's University, Kingston, 1975.

[228] L. Szpiro, *Lectures on equations defining space curves*, Tata Inst. of Fundamental Research, Bombay, 1979, and Berlin-Heidelberg-New York, (Springer), 1979.

IX,107. See Seminaire Bourbaki, nos. 453 and 458.

IX,108 to 111. See [177] (Seshadri) and [122].

IX,112 to 115. See Séminaire Bourbaki, no. 338, [177] (Seshadri), [188] and [197].

IX,116. See Séminaire Bourbaki, nos. 151 and 619.

IX,117. See Séminaire Bourbaki, nos. 93 and 145, [48] and:

[229] B. Mazur and W. Messing, *Universal extensions and One dimensional Crystalline cohomology*, Lecture Notes in Math., no. 370, Berlin-Heidelberg-New York (Springer), 1974.

IX,118. For the work of Picard, see [55]; for the work of Lefschetz and Severi see [56].

IX,119 to 121. See [177] (Messing).

IX,120. See:

[230] A. Weil, Sur la théorie des formes différentielles attachées à une variété analytique complexe, *Comm. Math. Helvet.*, 20 (1947), p. 110–116 (= *Oeuvres scientifiques*, vol. I, p. 374–380 (Springer), 1980).

IX,121. See:

[231] A. WEIL, Variétés abéliennes en Algèbre et théorie des nombres, *Coll. intern. CNRS*, 24 (Paris, 1949), p. 124–127. (= *Oeuvres scientifiques*, vol. I, p. 437–440 (Springer), 1980).

[232] J.-P. SERRE, Morphismes universels et différentielles de troisième espèce, *Séminaire Chevalley 1958–59: Variétés de Picard*, Exposé XI, Paris (Secr. math., 11, rue P.-Curie).

IX,123–126. See [177] (Mazur), [170] (Katz), and:

[233] A. WEIL, Number of solutions of equations in finite fields, *Bull. Amer. Math. Soc.*, 55 (1949), p. 497–508 (= *Oeuvres scientifiques*, vol. I, p. 399–410 (Springer), 1980).

[234] A. WEIL, Number theory and algebraic geometry, *Proc. of the Intern. Congress of mathematicians, Amsterdam 1954*, vol. III, p. 550–558, Amsterdam (North Holland), 1956 (= *Oeuvres scientifiques*, vol. II, p. 180–188 (Springer), 1980).

IX,127. See Séminaire Bourbaki, nos. 198 and 279 (or [221] (Grothendieck)). Up to now, the only complete proofs of formulas (69), (70) and (74) are to be found in [157].

IX,128. See [221] (Kleiman) for the "standard conjectures."

IX,129 to 139. See:

[235] P. DELIGNE, La conjecture de Weil. I, *Publ. math. de l'I.H.E.S.*, 43 (1974), p. 273–307.

For an account of the history of Deligne's proof, see [170] (Katz) and [177] (Mazur).

IX,140 to 145, see [177] (Messing) and:

[236] P. DELIGNE, La conjecture de Weil. II, *Publ. math. de l'I.H.E.S.*, 52 (1980), p. 137–252.

IX,146. See Séminaire Bourbaki no. 352 and [177] (Mazur). For p-divisible groups, see Séminaire Bourbaki, no. 318, and:

[237] M. DEMAZURE, *Lectures on p-divisible groups*, Lecture Notes in Math., no. 302, Berlin-Heidelberg-New York (Springer), 1972.

IX,147. For the proof of the analog to Mordell's conjecture for K a field of rational functions on a curve, see Séminaire Bourbaki no. 287.

IX,150. See Séminaire Bourbaki, no. 227.

IX,151 to 153. See Séminaire Bourbaki, no. 616. For the notion of height in Number theory, see Séminaire Bourbaki, no. 274.

IX,154. See Séminaire Bourbaki, no. 619.

IX,155.

1) See [184] (Iitaka).

2) See Séminaire Bourbaki, no. 537, and:

[238] *Géométrie algébrique réelle et formes quadratiques (Rennes, 1981)* (ed. J.-L. COLLIOT-THÉLÈNE, M. COSTE and L. MAHÉ), Lecture Notes in Math., no. 959, Berlin-Heidelberg-New York (Springer), 1982.

3) See Séminaire Bourbaki, no. 531.

4) See [177] (Katz, Ribet).

5) See Séminaire Bourbaki, no. 385, and:

[239] A. ASH, D. MUMFORD, M. RAPOPORT and Y. TAI, *Smooth compactification of locally symmetric varieties*, Brookline (Math. Sci. Press), 1975.

6) See [177] (Seshadri).

7) See [153] and [211].

8) See [221] (Grothendieck) and:

[240] *Brauer groups (Evanston, 1975)* (ed. D. ZELINSKY), Lecture Notes in Math., no. 549, Berlin-Heidelberg-New York (Springer), 1976.

[241] *Groupe de Brauer (Les Plans-sur-Bex, 1980)* (ed. M. KERVAIRE and M. OJANGUREN),

Lecture Notes in Math., no. 844, Berlin-Heidelberg-New York (Springer), 1981.

[242] *Brauer groups in ring theory and algebraic geometry (Antwerp, 1982)* (ed. F. VAN OYSTAYEN and A. VERSCHOREN), Lecture Notes in Math., no. 917, Berlin-Heidelberg-New York (Springer), 1982.

9) See Séminaire Bourbaki no. 365, [174] (Grothendieck), [175] (Kleiman) and:

[243] P. DELIGNE, J. MILNE, A. OGUS and K. SHIH, *Hodge cycles, motives and Shimura varieties*, Lecture Notes in Math., no. 900, Berlin-Heidelberg-New York (Springer), 1982.

10) See Séminaire Bourbaki nos. 585 and 589, and [201] (J. Cheeger, M. Goresky and R. MacPherson).

11) See:

[244] M. ARTIN and B. MAZUR, *Etale homotopy*, Lecture Notes in Math., no. 100, Berlin-Heidelberg-New York (Springer), 1969.

12) See [168] (Clemens), [176] (D. Mumford), [177] (Hartshorne) and:

[245] A. BEAUVILLE, Variétés de Prym et jacobiennes intermédiaires, *Ann. Ecole Norm. Sup.* (4), 10 (1977), p. 309–391.

13) See Séminaire Bourbaki no. 548.

Recent bibliography;

[246] M. MIYANISHI, *Non complete algebraic surfaces*, Lecture Notes in Math., no. 857, Berlin-Heidelberg-New York (Springer), 1981.

[247] B. ANGÉNIOL, *Familles de cycles algébriques*, Lecture Notes in Math., no. 896, Berlin-Heidelberg-New York (Springer), 1981.

[248] *Enumerative geometry and classical algebraic geometry* (ed. P. LE BARZ and Y. HERVIER), Progress in math., no. 24, Boston (Birkhäuser), 1982.

[249] *Algebraic threefolds (C.I.M.E., Varenna, 1981)* (ed. A. CONTE), Lecture Notes in Math., no. 947, Berlin-Heidelberg-New York (Springer), 1982.

[250] Séminaire sur les pinceaux de courbes de genre au moins deux (L. SZPIRO), *Astérisque*, 86 (1981).

[251] *Algebraic geometry: open problems (Ravello, 1982)* (ed. C. CILIBERTO, F. GHIONE, F. ORECCHIA), Lecture Notes in Math., no. 997, Berlin-Heidelberg-New York (Springer), 1983.

[252] *Algebraic geometry (Ann Arbor, 1981)* (ed. J. DOLGACHEV), Lecture Notes in Math., no. 1008, Berlin-Heidelberg-New York (Springer), 1983.

[253] *Algebraic geometry (Tokyo/Kyoto, 1982)* (ed. M. RAYNAUD and T. SHIODA), Lecture Notes in Math., no. 1016, Berlin-Heidelberg-New York (Springer), 1983.

The Bourbaki Seminar "Exposés" have been published:

from no. 1 to no. 346 by W. A. Benjamin, Inc., New York;

from no. 347 to no. 578 in Lecture Notes in Math., nos. 179, 180, 244, 317, 383, 431, 514, 567, 677, Berlin-Heidelberg-New York (Springer);

from no. 579 to 614 in *Astérisque*, 92–93 (1982) and 105–106 (1983).

Index of Cited Names[1]

1. The roman numerals refer to chapters and the arabic numerals refer to the paragraphs in each chapter.

Index of Terminology